TINYWOW

MATH MASTERY WORKBOOK

GRADE **3**

100 DAYS CHALLENGE

ADDITION
SUBTRACTION
MULTIPLICATION
DIVISION
FRACTIONS

Math Mastery Workbook provides:

- 100 math mastery tests
- 30 math practice problems per page
- Answer key is included in the back of the book
- Progress sheet
- Math award

Practice tests
can be used for:

- Homework
- Home-school
- Tutoring

This workbook is intended for students who have already studied addition and subtraction.

IMPROVE YOUR CHILD'S SUCCESS IN CLASS

MORE BOOKS

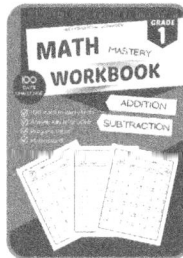

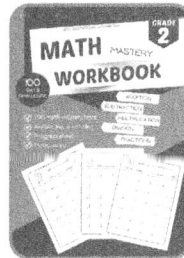

www.findoutandlearn.com

Hello and Welcome to NESY

We are a small publishing company started by creative parents who felt inspired by their kids' joy of discovery and learning.
We believe that learning math is a magical experience that opens doors to critical thinking and knowledge.

Our mission is to create workbooks for kids that they can use by themselves or with their parents and teachers.

Learning should be fun, so our workbooks are designed to be enjoyable and full of practices that help your brain grow.

Explore more books
and look for more activities online
at: www.findoutandlearn.com

Easy-to-use
and Free resources.
Scan the QR code below

Visit: www.findoutandlearn.com

We are offering
free printable worksheets,
educational tasks, fun crafts, and
kid`s activities.

This book
belongs to:

$\frac{1}{4}$

A+

INTRODUCTION

The first step to developing conceptual understanding in math is for students to recall addition, subtraction, multiplication, division, and fraction facts. This math fact fluency brings on automaticity, and when students achieve automaticity of their facts a common benchmark is that they can recall thirty problems in one minute. This skill is essential for students to master and will impact their future learning of more difficult mathematics concepts.

This workbook offers daily facts fluency practice in the form of written one-minute time tests. These are done every day. These tests are created so that teachers can rapidly assess students' progress and for students to get written practice to develop accuracy and speed in addition, subtraction, multiplication, division, and fraction. With the progress check sheet included, students also get a chance to keep track of their progress.

Best results are obtained only after teachers have explicitly taught strategies, provided opportunities for the use of manipulatives and students have gotten multiple opportunities to practice the strategies.

TABLE OF CONTENT

Addition

1) 94
  + 93

2) 87
  + 76

3) 86
  + 75

4) 86
  + 95

5) 90
  + 80

6) 95
  + 95

7) 94
  + 91

8) 86
  + 81

9) 96
  + 86

10) 92
   + 90

11) 97
   + 86

12) 96
   + 77

13) 85
   + 86

14) 95
   + 97

15) 97
   + 77

16) 98
   + 84

17) 93
   + 84

18) 92
   + 81

19) 95
   + 76

20) 95
   + 77

21) 81
   + 98

22) 96
   + 96

23) 93
   + 90

24) 91
   + 87

25) 84
   + 98

26) 82
   + 92

27) 80
   + 85

28) 82
   + 89

29) 99
   + 79

30) 81
   + 76

1)   98
   + 87

2)   91
   + 93

3)   87
   + 80

4)   94
   + 75

5)   92
   + 80

6)   80
   + 96

7)   87
   + 90

8)   84
   + 83

9)   97
   + 77

10)   89
    + 96

11)   96
    + 90

12)   93
    + 91

13)   82
    + 96

14)   80
    + 94

15)   82
    + 91

16)   96
    + 97

17)   94
    + 98

18)   94
    + 93

19)   90
    + 84

20)   96
    + 78

21)   97
    + 90

22)   85
    + 95

23)   84
    + 91

24)   98
    + 98

25)   90
    + 76

26)   89
    + 77

27)   91
    + 79

28)   91
    + 84

29)   92
    + 87

30)   91
    + 82

1)
$$\begin{array}{r} 102 \\ +\ \ 96 \\ \hline \end{array}$$

2)
$$\begin{array}{r} 87 \\ +\ 98 \\ \hline \end{array}$$

3)
$$\begin{array}{r} 96 \\ +\ 95 \\ \hline \end{array}$$

4)
$$\begin{array}{r} 91 \\ +\ 97 \\ \hline \end{array}$$

5)
$$\begin{array}{r} 93 \\ +\ 90 \\ \hline \end{array}$$

6)
$$\begin{array}{r} 105 \\ +\ \ 93 \\ \hline \end{array}$$

7)
$$\begin{array}{r} 109 \\ +\ \ 87 \\ \hline \end{array}$$

8)
$$\begin{array}{r} 80 \\ +\ 82 \\ \hline \end{array}$$

9)
$$\begin{array}{r} 82 \\ +\ 91 \\ \hline \end{array}$$

10)
$$\begin{array}{r} 108 \\ +\ \ 92 \\ \hline \end{array}$$

11)
$$\begin{array}{r} 95 \\ +\ 98 \\ \hline \end{array}$$

12)
$$\begin{array}{r} 92 \\ +\ 86 \\ \hline \end{array}$$

13)
$$\begin{array}{r} 82 \\ +\ 98 \\ \hline \end{array}$$

14)
$$\begin{array}{r} 109 \\ +\ \ 83 \\ \hline \end{array}$$

15)
$$\begin{array}{r} 84 \\ +\ 82 \\ \hline \end{array}$$

16)
$$\begin{array}{r} 92 \\ +\ 91 \\ \hline \end{array}$$

17)
$$\begin{array}{r} 103 \\ +\ \ 84 \\ \hline \end{array}$$

18)
$$\begin{array}{r} 103 \\ +\ \ 88 \\ \hline \end{array}$$

19)
$$\begin{array}{r} 98 \\ +\ 85 \\ \hline \end{array}$$

20)
$$\begin{array}{r} 81 \\ +\ 86 \\ \hline \end{array}$$

21)
$$\begin{array}{r} 104 \\ +\ \ 88 \\ \hline \end{array}$$

22)
$$\begin{array}{r} 86 \\ +\ 91 \\ \hline \end{array}$$

23)
$$\begin{array}{r} 106 \\ +\ \ 87 \\ \hline \end{array}$$

24)
$$\begin{array}{r} 94 \\ +\ 97 \\ \hline \end{array}$$

25)
$$\begin{array}{r} 109 \\ +\ \ 93 \\ \hline \end{array}$$

26)
$$\begin{array}{r} 82 \\ +\ 82 \\ \hline \end{array}$$

27)
$$\begin{array}{r} 93 \\ +\ 97 \\ \hline \end{array}$$

28)
$$\begin{array}{r} 101 \\ +\ \ 95 \\ \hline \end{array}$$

29)
$$\begin{array}{r} 83 \\ +\ 82 \\ \hline \end{array}$$

30)
$$\begin{array}{r} 93 \\ +\ 89 \\ \hline \end{array}$$

1)
```
   107
 +  91
```

2)
```
   120
 +  87
```

3)
```
   105
 +  91
```

4)
```
    81
 +  87
```

5)
```
    81
 +  98
```

6)
```
    89
 +  95
```

7)
```
   117
 +  86
```

8)
```
   106
 +  87
```

9)
```
   117
 +  92
```

10)
```
   119
 +  98
```

11)
```
    90
 +  92
```

12)
```
    81
 +  88
```

13)
```
    85
 +  86
```

14)
```
    84
 +  96
```

15)
```
    93
 +  93
```

16)
```
    87
 +  88
```

17)
```
   109
 +  97
```

18)
```
    97
 +  86
```

19)
```
    86
 +  96
```

20)
```
   103
 +  90
```

21)
```
   120
 +  98
```

22)
```
    80
 +  91
```

23)
```
   118
 +  92
```

24)
```
    99
 +  87
```

25)
```
   117
 +  88
```

26)
```
    95
 +  90
```

27)
```
   100
 +  93
```

28)
```
   113
 +  92
```

29)
```
    83
 +  89
```

30)
```
    87
 +  88
```

1)
$$104 + 98$$

2)
$$96 + 94$$

3)
$$99 + 101$$

4)
$$132 + 104$$

5)
$$101 + 100$$

6)
$$98 + 98$$

7)
$$134 + 108$$

8)
$$104 + 99$$

9)
$$117 + 99$$

10)
$$129 + 100$$

11)
$$104 + 97$$

12)
$$120 + 94$$

13)
$$131 + 103$$

14)
$$99 + 98$$

15)
$$124 + 99$$

16)
$$122 + 94$$

17)
$$110 + 99$$

18)
$$127 + 94$$

19)
$$126 + 93$$

20)
$$108 + 96$$

21)
$$119 + 93$$

22)
$$111 + 107$$

23)
$$105 + 99$$

24)
$$121 + 106$$

25)
$$124 + 102$$

26)
$$107 + 106$$

27)
$$122 + 106$$

28)
$$99 + 95$$

29)
$$109 + 94$$

30)
$$112 + 108$$

1) 100
  + 88
  [ ]

2)   96
  + 91
  [ ]

3)   95
  + 144
  [ ]

4) 115
  + 137
  [ ]

5)   89
  + 95
  [ ]

6) 122
  + 126
  [ ]

7) 125
  + 94
  [ ]

8)   85
  + 112
  [ ]

9)   91
  + 101
  [ ]

10) 128
  + 113
  [ ]

11) 106
  + 90
  [ ]

12) 101
  + 145
  [ ]

13) 101
  + 101
  [ ]

14)   87
  + 128
  [ ]

15) 105
  + 108
  [ ]

16) 107
  + 129
  [ ]

17)   97
  + 104
  [ ]

18)   95
  + 107
  [ ]

19) 109
  + 108
  [ ]

20)   82
  + 94
  [ ]

21)   91
  + 115
  [ ]

22) 114
  + 137
  [ ]

23) 101
  + 98
  [ ]

24)   92
  + 96
  [ ]

25)   86
  + 119
  [ ]

26) 100
  + 91
  [ ]

27) 130
  + 118
  [ ]

28) 122
  + 118
  [ ]

29) 103
  + 135
  [ ]

30) 122
  + 136
  [ ]

1)
```
   103
 + 131
 _____
```

2)
```
   114
 +  91
 _____
```

3)
```
   126
 + 105
 _____
```

4)
```
    94
 + 121
 _____
```

5)
```
   117
 +  99
 _____
```

6)
```
   104
 + 127
 _____
```

7)
```
   115
 + 144
 _____
```

8)
```
   100
 +  92
 _____
```

9)
```
    80
 +  97
 _____
```

10)
```
   102
 + 141
 _____
```

11)
```
   121
 + 126
 _____
```

12)
```
   124
 + 141
 _____
```

13)
```
   102
 +  95
 _____
```

14)
```
    83
 + 127
 _____
```

15)
```
   108
 +  90
 _____
```

16)
```
    96
 + 105
 _____
```

17)
```
   128
 +  89
 _____
```

18)
```
   130
 + 144
 _____
```

19)
```
    84
 +  93
 _____
```

20)
```
    85
 +  90
 _____
```

21)
```
    98
 + 126
 _____
```

22)
```
    80
 + 112
 _____
```

23)
```
   100
 +  99
 _____
```

24)
```
    97
 + 131
 _____
```

25)
```
    95
 + 119
 _____
```

26)
```
   105
 + 144
 _____
```

27)
```
   104
 + 122
 _____
```

28)
```
    90
 + 114
 _____
```

29)
```
    98
 + 122
 _____
```

30)
```
   125
 + 103
 _____
```

1)
```
  189
+ 157
```

2)
```
  242
+ 257
```

3)
```
  253
+ 222
```

4)
```
  205
+ 189
```

5)
```
  234
+ 144
```

6)
```
  197
+ 219
```

7)
```
  206
+ 169
```

8)
```
  199
+ 261
```

9)
```
  236
+ 183
```

10)
```
  194
+ 215
```

11)
```
  154
+ 173
```

12)
```
  241
+ 196
```

13)
```
  165
+ 147
```

14)
```
  232
+ 216
```

15)
```
  224
+ 198
```

16)
```
  193
+ 142
```

17)
```
  228
+ 165
```

18)
```
  147
+ 226
```

19)
```
  207
+ 173
```

20)
```
  181
+ 180
```

21)
```
  224
+ 233
```

22)
```
  239
+ 156
```

23)
```
  163
+ 184
```

24)
```
  218
+ 249
```

25)
```
  232
+ 204
```

26)
```
  243
+ 170
```

27)
```
  219
+ 151
```

28)
```
  146
+ 253
```

29)
```
  205
+ 207
```

30)
```
  147
+ 232
```

1) 106
  + 149

2) 119
  + 133

3) 119
  + 113

4) 167
  + 171

5) 176
  + 141

6) 130
  + 102

7) 185
  + 176

8) 140
  + 94

9) 169
  + 186

10) 125
   + 170

11) 132
   + 169

12) 174
   + 95

13)   96
   + 100

14) 103
   + 134

15)   96
   + 174

16) 162
   + 180

17)   96
   + 92

18) 149
   + 121

19) 185
   + 172

20) 150
   + 112

21) 145
   + 174

22)   87
   + 180

23)   90
   + 99

24) 163
   + 100

25) 122
   + 139

26) 134
   + 127

27) 106
   + 131

28) 179
   + 99

29) 109
   + 152

30) 154
   + 97

1) 144
  + 167
  ☐

2)   98
  + 133
  ☐

3)   80
  + 133
  ☐

4) 203
  + 217
  ☐

5) 134
  + 177
  ☐

6)   90
  + 160
  ☐

7) 148
  + 168
  ☐

8) 130
  + 145
  ☐

9)   97
  + 139
  ☐

10)   87
  + 121
  ☐

11) 111
  + 201
  ☐

12) 148
  + 100
  ☐

13) 118
  + 216
  ☐

14) 119
  + 100
  ☐

15) 185
  + 179
  ☐

16) 209
  + 101
  ☐

17) 133
  + 126
  ☐

18)   95
  + 152
  ☐

19)   99
  + 136
  ☐

20) 154
  + 219
  ☐

21) 202
  + 175
  ☐

22) 162
  + 194
  ☐

23) 145
  + 138
  ☐

24) 158
  + 198
  ☐

25)   81
  + 109
  ☐

26) 182
  + 138
  ☐

27)   90
  + 138
  ☐

28) 104
  + 220
  ☐

29) 154
  + 109
  ☐

30) 170
  + 169
  ☐

1)   159
    + 163
    [_____]

2)   198
    + 217
    [_____]

3)   148
    + 140
    [_____]

4)   179
    + 149
    [_____]

5)   171
    + 115
    [_____]

6)   182
    + 173
    [_____]

7)   168
    + 120
    [_____]

8)   149
    + 194
    [_____]

9)   197
    + 149
    [_____]

10)   114
    + 144
    [_____]

11)   121
    + 157
    [_____]

12)   188
    + 181
    [_____]

13)   103
    + 139
    [_____]

14)   197
    + 212
    [_____]

15)   129
    + 205
    [_____]

16)   114
    + 198
    [_____]

17)   184
    + 209
    [_____]

18)   107
    + 150
    [_____]

19)   120
    + 177
    [_____]

20)   132
    + 164
    [_____]

21)   170
    ı 137
    [_____]

22)   147
    + 150
    [_____]

23)   157
    + 142
    [_____]

24)   139
    + 189
    [_____]

25)   199
    + 162
    [_____]

26)   183
    + 180
    [_____]

27)   121
    + 210
    [_____]

28)   165
    + 168
    [_____]

29)   104
    + 198
    [_____]

30)   150
    + 143
    [_____]

1) 183
  + 180

2) 151
  + 229

3) 180
  + 250

4) 214
  + 189

5) 197
  + 118

6) 190
  + 239

7) 187
  + 179

8) 169
  + 207

9) 163
  + 161

10) 134
   + 171

11) 198
   + 247

12) 218
   + 242

13) 132
   + 127

14) 142
   + 158

15) 182
   + 217

16) 208
   + 116

17) 122
   + 114

18) 178
   + 230

19) 218
   + 129

20) 207
   + 232

21) 197
   + 117

22) 149
   + 164

23) 214
   + 244

24) 228
   + 170

25) 156
   + 163

26) 148
   + 163

27) 152
   + 201

28) 175
   + 155

29) 162
   + 148

30) 211
   + 229

1)
```
   123
 + 202
```

2)
```
   135
 + 165
```

3)
```
   197
 + 212
```

4)
```
   138
 + 125
```

5)
```
   164
 + 191
```

6)
```
   190
 + 173
```

7)
```
   183
 + 200
```

8)
```
   199
 + 193
```

9)
```
   197
 + 178
```

10)
```
   182
 + 137
```

11)
```
   151
 + 184
```

12)
```
   105
 + 210
```

13)
```
   138
 + 114
```

14)
```
   134
 + 177
```

15)
```
   131
 + 161
```

16)
```
   124
 + 208
```

17)
```
   145
 + 143
```

18)
```
   114
 + 210
```

19)
```
   133
 + 205
```

20)
```
   128
 + 169
```

21)
```
   133
 + 213
```

22)
```
   132
 + 114
```

23)
```
   196
 + 112
```

24)
```
   111
 + 130
```

25)
```
   175
 + 137
```

26)
```
   123
 + 142
```

27)
```
   183
 + 173
```

28)
```
   123
 + 126
```

29)
```
   184
 + 125
```

30)
```
   177
 + 124
```

1)
```
   181
 + 125
```

2)
```
   113
 + 151
```

3)
```
   131
 + 206
```

4)
```
   218
 + 119
```

5)
```
   138
 + 168
```

6)
```
   162
 + 170
```

7)
```
   173
 + 167
```

8)
```
   117
 + 172
```

9)
```
   153
 + 214
```

10)
```
   116
 + 210
```

11)
```
   111
 + 127
```

12)
```
   227
 + 114
```

13)
```
   214
 + 213
```

14)
```
   103
 + 115
```

15)
```
   164
 + 133
```

16)
```
   109
 + 142
```

17)
```
   168
 + 187
```

18)
```
   192
 + 178
```

19)
```
   162
 + 162
```

20)
```
   198
 + 115
```

21)
```
   178
 + 164
```

22)
```
   149
 + 207
```

23)
```
   174
 + 160
```

24)
```
   196
 + 182
```

25)
```
   142
 + 209
```

26)
```
   208
 + 150
```

27)
```
   160
 + 142
```

28)
```
   184
 + 112
```

29)
```
   133
 + 204
```

30)
```
   185
 + 120
```

1)
```
  436
+ 199
------
```

2)
```
  318
+ 201
------
```

3)
```
  294
+ 296
------
```

4)
```
  308
+ 295
------
```

5)
```
  396
+ 218
------
```

6)
```
  282
+ 269
------
```

7)
```
  432
+ 293
------
```

8)
```
  321
+ 243
------
```

9)
```
  337
+ 304
------
```

10)
```
  356
+ 246
------
```

11)
```
  353
+ 175
------
```

12)
```
  259
+ 296
------
```

13)
```
  427
+ 198
------
```

14)
```
  282
+ 316
------
```

15)
```
  351
+ 285
------
```

16)
```
  407
+ 219
------
```

17)
```
  327
+ 269
------
```

18)
```
  371
+ 200
------
```

19)
```
  317
+ 238
------
```

20)
```
  291
+ 229
------
```

21)
```
  358
+ 178
------
```

22)
```
  363
+ 255
------
```

23)
```
  317
+ 201
------
```

24)
```
  219
+ 250
------
```

25)
```
  315
+ 193
------
```

26)
```
  300
+ 188
------
```

27)
```
  239
+ 180
------
```

28)
```
  273
+ 246
------
```

29)
```
  415
+ 175
------
```

30)
```
  311
+ 297
------
```

1) 92 + 97 = [   ]   2) 88 + 77 = [   ]   3) 99 + 132 = [   ]

4) 94 + 128 = [   ]   5) 86 + 120 = [   ]   6) 99 + 79 = [   ]

7) 87 + 95 = [   ]   8) 92 + 88 = [   ]   9) 92 + 88 = [   ]

10) 88 + 117 = [   ]   11) 94 + 133 = [   ]   12) 90 + 109 = [   ]

13) 87 + 78 = [   ]   14) 86 + 120 = [   ]   15) 91 + 99 = [   ]

16) 86 + 86 = [   ]   17) 96 + 112 = [   ]   18) 94 + 79 = [   ]

19) 90 + 143 = [   ]   20) 99 + 116 = [   ]   21) 90 + 108 = [   ]

22) 95 + 144 = [   ]   23) 99 + 144 = [   ]   24) 85 + 77 = [   ]

25) 94 + 111 = [   ]   26) 92 + 91 = [   ]   27) 97 + 77 = [   ]

28) 95 + 94 = [   ]   29) 95 + 124 = [   ]   30) 88 + 80 = [   ]

1) $90 + 139 =$ ☐   2) $99 + 115 =$ ☐   3) $89 + 89 =$ ☐

4) $92 + 79 =$ ☐   5) $85 + 98 =$ ☐   6) $94 + 147 =$ ☐

7) $85 + 123 =$ ☐   8) $97 + 105 =$ ☐   9) $96 + 125 =$ ☐

10) $99 + 128 =$ ☐   11) $96 + 90 =$ ☐   12) $91 + 93 =$ ☐

13) $88 + 118 =$ ☐   14) $89 + 141 =$ ☐   15) $92 + 90 =$ ☐

16) $95 + 101 =$ ☐   17) $90 + 128 =$ ☐   18) $93 + 110 =$ ☐

19) $96 + 109 =$ ☐   20) $90 + 77 =$ ☐   21) $97 + 108 =$ ☐

22) $99 + 83 =$ ☐   23) $88 + 123 =$ ☐   24) $85 + 104 =$ ☐

25) $99 + 119 =$ ☐   26) $86 + 134 =$ ☐   27) $89 + 126 =$ ☐

28) $99 + 93 =$ ☐   29) $99 + 88 =$ ☐   30) $97 + 146 =$ ☐

1) 94 + 133 = ☐   2) 89 + 83 = ☐   3) 88 + 84 = ☐

4) 88 + 143 = ☐   5) 89 + 81 = ☐   6) 95 + 118 = ☐

7) 98 + 80 = ☐   8) 97 + 121 = ☐   9) 91 + 108 = ☐

10) 98 + 96 = ☐   11) 89 + 114 = ☐   12) 99 + 111 = ☐

13) 86 + 82 = ☐   14) 86 + 102 = ☐   15) 91 + 107 = ☐

16) 88 + 88 = ☐   17) 94 + 131 = ☐   18) 89 + 134 = ☐

19) 91 + 127 = ☐   20) 98 + 120 = ☐   21) 97 + 113 = ☐

22) 98 + 80 = ☐   23) 97 + 128 = ☐   24) 92 + 123 = ☐

25) 96 + 103 = ☐   26) 94 + 111 = ☐   27) 93 + 101 = ☐

28) 99 + 100 = ☐   29) 97 + 101 = ☐   30) 90 + 107 = ☐

1) $153 + 109 =$ ☐ 2) $143 + 200 =$ ☐ 3) $103 + 84 =$ ☐

4) $155 + 116 =$ ☐ 5) $204 + 169 =$ ☐ 6) $163 + 99 =$ ☐

7) $124 + 118 =$ ☐ 8) $193 + 159 =$ ☐ 9) $163 + 193 =$ ☐

10) $156 + 125 =$ ☐ 11) $169 + 142 =$ ☐ 12) $120 + 112 =$ ☐

13) $99 + 129 =$ ☐ 14) $201 + 184 =$ ☐ 15) $189 + 163 =$ ☐

16) $182 + 188 =$ ☐ 17) $122 + 143 =$ ☐ 18) $153 + 178 =$ ☐

19) $201 + 107 =$ ☐ 20) $183 + 167 =$ ☐ 21) $135 + 145 =$ ☐

22) $98 + 174 =$ ☐ 23) $179 + 135 =$ ☐ 24) $132 + 119 =$ ☐

25) $131 + 109 =$ ☐ 26) $135 + 98 =$ ☐ 27) $189 + 181 =$ ☐

28) $129 + 122 =$ ☐ 29) $173 + 190 =$ ☐ 30) $138 + 190 =$ ☐

1) $145 + 200 =$ ☐   2) $195 + 139 =$ ☐   3) $223 + 233 =$ ☐

4) $267 + 151 =$ ☐   5) $159 + 96 =$ ☐   6) $264 + 90 =$ ☐

7) $257 + 88 =$ ☐   8) $240 + 197 =$ ☐   9) $191 + 233 =$ ☐

10) $273 + 162 =$ ☐   11) $215 + 111 =$ ☐   12) $273 + 214 =$ ☐

13) $158 + 180 =$ ☐   14) $133 + 237 =$ ☐   15) $128 + 141 =$ ☐

16) $122 + 107 =$ ☐   17) $212 + 117 =$ ☐   18) $281 + 217 =$ ☐

19) $293 + 214 =$ ☐   20) $116 + 206 =$ ☐   21) $155 + 131 =$ ☐

22) $254 + 212 =$ ☐   23) $167 + 232 =$ ☐   24) $165 + 229 =$ ☐

25) $278 + 147 =$ ☐   26) $258 + 201 =$ ☐   27) $256 + 125 =$ ☐

28) $297 + 202 =$ ☐   29) $200 + 210 =$ ☐   30) $211 + 191 =$ ☐

Subtraction

1) $87 - 50 =$ 2) $65 - 61 =$ 3) $84 - 80 =$

4) $76 - 62 =$ 5) $79 - 63 =$ 6) $85 - 81 =$

7) $84 - 55 =$ 8) $73 - 46 =$ 9) $50 - 47 =$

10) $90 - 63 =$ 11) $65 - 44 =$ 12) $86 - 40 =$

13) $84 - 40 =$ 14) $68 - 52 =$ 15) $70 - 53 =$

16) $75 - 52 =$ 17) $69 - 43 =$ 18) $68 - 63 =$

19) $78 - 44 =$ 20) $81 - 54 =$ 21) $89 - 84 =$

22) $70 - 41 =$ 23) $86 - 55 =$ 24) $89 - 51 =$

25) $87 - 43 =$ 26) $90 - 78 =$ 27) $75 - 50 =$

28) $80 - 40 =$ 29) $71 - 49 =$ 30) $83 - 42 =$

1) $52 - 43 =$ 2) $56 - 46 =$ 3) $78 - 67 =$

4) $87 - 76 =$ 5) $72 - 59 =$ 6) $83 - 81 =$

7) $73 - 58 =$ 8) $63 - 50 =$ 9) $68 - 54 =$

10) $67 - 60 =$ 11) $49 - 43 =$ 12) $78 - 76 =$

13) $74 - 56 =$ 14) $80 - 71 =$ 15) $81 - 41 =$

16) $64 - 41 =$ 17) $76 - 76 =$ 18) $54 - 51 =$

19) $86 - 86 =$ 20) $56 - 45 =$ 21) $86 - 70 =$

22) $55 - 52 =$ 23) $62 - 10 =$ 24) $86 - 85 =$

25) $90 - 76 =$ 26) $80 - 59 =$ 27) $87 - 51 =$

28) $87 - 83 =$ 29) $78 - 44 =$ 30) $58 - 52 =$

1) 63 - 61 =   2) 93 - 84 =   3) 95 - 80 =

4) 71 - 68 =   5) 87 - 81 =   6) 74 - 70 =

7) 70 - 62 =   8) 75 - 74 =   9) 82 - 55 =

10) 69 - 56 =   11) 82 - 57 =   12) 88 - 85 =

13) 84 - 61 =   14) 88 - 50 =   15) 74 - 58 =

16) 95 - 61 =   17) 64 - 56 =   18) 92 - 67 =

19) 60 - 59 =   20) 63 - 63 =   21) 77 - 61 =

22) 66 - 58 =   23) 92 - 51 =   24) 95 - 84 =

25) 87 - 83 =   26) 69 - 54 =   27) 93 - 79 =

28) 90 - 51 =   29) 77 - 75 =   30) 89 - 63 =

1) 93 - 73 =   2) 84 - 74 =   3) 100 - 82 =

4) 75 - 75 =   5) 107 - 77 =   6) 100 - 91 =

7) 96 - 81 =   8) 94 - 75 =   9) 89 - 88 =

10) 103 - 96 =   11) 102 - 75 =   12) 104 - 93 =

13) 108 - 83 =   14) 109 - 80 =   15) 100 - 94 =

16) 105 - 85 =   17) 110 - 85 =   18) 93 - 88 =

19) 103 - 69 =   20) 103 - 91 =   21) 102 - 77 =

22) 108 - 90 =   23) 87 - 69 =   24) 93 - 74 =

25) 81 - 77 =   26) 92 - 73 =   27) 108 - 90 =

28) 88 - 88 =   29) 87 - 80 =   30) 92 - 73 =

1) $104 - 74 =$ 2) $92 - 74 =$ 3) $101 - 78 =$

4) $109 - 88 =$ 5) $88 - 85 =$ 6) $113 - 85 =$

7) $111 - 96 =$ 8) $96 - 89 =$ 9) $110 - 78 =$

10) $83 - 74 =$ 11) $104 - 70 =$ 12) $115 - 94 =$

13) $106 - 74 =$ 14) $99 - 83 =$ 15) $104 - 97 =$

16) $102 - 83 =$ 17) $94 - 76 =$ 18) $82 - 78 =$

19) $85 - 81 =$ 20) $87 - 71 =$ 21) $108 - 73 =$

22) $102 - 96 =$ 23) $102 - 92 =$ 24) $96 - 88 =$

25) $78 - 73 =$ 26) $111 - 76 =$ 27) $97 - 74 =$

28) $113 - 74 =$ 29) $94 - 84 =$ 30) $109 - 95 =$

1)
```
  125
-  98
```

2)
```
  150
-  91
```

3)
```
  150
- 139
```

4)
```
  149
-  90
```

5)
```
  120
-  97
```

6)
```
   98
-  94
```

7)
```
  101
-  95
```

8)
```
  130
- 128
```

9)
```
  150
- 113
```

10)
```
  135
- 126
```

11)
```
  149
-  89
```

12)
```
   98
-  91
```

13)
```
  145
- 110
```

14)
```
  126
- 121
```

15)
```
  129
-  98
```

16)
```
  148
- 132
```

17)
```
  141
- 109
```

18)
```
  146
-  98
```

19)
```
  122
-  89
```

20)
```
  119
-  98
```

21)
```
  148
- 129
```

22)
```
  138
- 102
```

23)
```
  127
- 122
```

24)
```
  129
- 114
```

25)
```
  126
-  92
```

26)
```
  146
- 138
```

27)
```
  102
-  98
```

28)
```
   92
-  90
```

29)
```
  109
- 103
```

30)
```
  122
-  96
```

1) 128
   − 119
   _____

2) 195
   − 115
   _____

3) 211
   − 170
   _____

4) 119
   − 105
   _____

5) 182
   − 151
   _____

6) 148
   − 125
   _____

7) 153
   − 127
   _____

8) 186
   − 160
   _____

9) 159
   − 124
   _____

10) 206
    − 105
    _____

11) 187
    − 187
    _____

12) 195
    − 195
    _____

13) 179
    − 139
    _____

14) 181
    − 126
    _____

15) 213
    − 147
    _____

16) 152
    − 121
    _____

17) 217
    − 190
    _____

18) 189
    − 114
    _____

19) 157
    − 107
    _____

20) 216
    − 178
    _____

21) 168
    − 147
    _____

22) 154
    − 152
    _____

23) 204
    − 189
    _____

24) 162
    − 110
    _____

25) 176
    − 104
    _____

26) 200
    − 159
    _____

27) 185
    − 132
    _____

28) 197
    − 153
    _____

29) 197
    − 172
    _____

30) 215
    − 195
    _____

1)   196
   - 190
   ☐

2)   111
   - 100
   ☐

3)   221
   - 146
   ☐

4)   156
   - 113
   ☐

5)   213
   - 189
   ☐

6)   135
   - 107
   ☐

7)   203
   - 127
   ☐

8)   201
   - 163
   ☐

9)   211
   - 210
   ☐

10)   205
   - 175
   ☐

11)   159
   - 157
   ☐

12)   216
   - 170
   ☐

13)   172
   - 158
   ☐

14)   157
   - 118
   ☐

15)   113
   - 112
   ☐

16)   156
   - 137
   ☐

17)   163
   - 119
   ☐

18)   110
   - 106
   ☐

19)   160
   - 105
   ☐

20)   189
   - 160
   ☐

21)   205
   - 130
   ☐

22)   111
   - 106
   ☐

23)   131
   - 114
   ☐

24)   150
   - 98
   ☐

25)   133
   - 112
   ☐

26)   204
   - 101
   ☐

27)   154
   - 118
   ☐

28)   168
   - 121
   ☐

29)   211
   - 142
   ☐

30)   162
   - 123
   ☐

NAME:--------DATE:--------

DAY:29   TIME:--------- SCORE: /30

1) 213 − 165

2) 242 − 121

3) 238 − 145

4) 171 − 137

5) 195 − 114

6) 161 − 117

7) 204 − 114

8) 248 − 170

9) 248 − 117

10) 125 − 101

11) 189 − 172

12) 181 − 164

13) 153 − 132

14) 246 − 152

15) 226 − 225

16) 118 − 108

17) 190 − 146

18) 212 − 139

19) 174 − 156

20) 159 − 101

21) 210 − 132

22) 231 − 104

23) 222 − 204

24) 199 − 136

25) 203 − 121

26) 220 − 199

27) 210 − 109

28) 143 − 111

29) 221 − 174

30) 189 − 129

37

1)
$$217 - 187$$

2)
$$218 - 112$$

3)
$$188 - 107$$

4)
$$270 - 122$$

5)
$$242 - 123$$

6)
$$257 - 231$$

7)
$$215 - 124$$

8)
$$261 - 194$$

9)
$$258 - 148$$

10)
$$232 - 109$$

11)
$$196 - 119$$

12)
$$208 - 159$$

13)
$$169 - 124$$

14)
$$191 - 153$$

15)
$$154 - 129$$

16)
$$167 - 152$$

17)
$$243 - 139$$

18)
$$263 - 205$$

19)
$$226 - 203$$

20)
$$214 - 202$$

21)
$$244 - 172$$

22)
$$171 - 147$$

23)
$$254 - 110$$

24)
$$194 - 105$$

25)
$$187 - 147$$

26)
$$237 - 172$$

27)
$$189 - 153$$

28)
$$124 - 119$$

29)
$$238 - 219$$

30)
$$242 - 148$$

1)
```
   274
-  221
_____
```

2)
```
   233
-  195
_____
```

3)
```
   320
-  155
_____
```

4)
```
   302
-  167
_____
```

5)
```
   265
-  199
_____
```

6)
```
   324
-  100
_____
```

7)
```
   288
-  109
_____
```

8)
```
   290
-  185
_____
```

9)
```
   230
-  117
_____
```

10)
```
   257
-  240
_____
```

11)
```
   298
-  271
_____
```

12)
```
   292
-  234
_____
```

13)
```
   267
-  245
_____
```

14)
```
   278
-  176
_____
```

15)
```
   314
-  302
_____
```

16)
```
   164
-  139
_____
```

17)
```
   230
-  102
_____
```

18)
```
   278
-  201
_____
```

19)
```
   238
-  206
_____
```

20)
```
   260
-  198
_____
```

21)
```
   296
-   97
_____
```

22)
```
   246
-  152
_____
```

23)
```
   330
-  218
_____
```

24)
```
   175
-  115
_____
```

25)
```
   294
-  105
_____
```

26)
```
   159
-  119
_____
```

27)
```
   288
-  178
_____
```

28)
```
   247
-  126
_____
```

29)
```
   189
-  111
_____
```

30)
```
   234
-  125
_____
```

1) 258 − 222

2) 319 − 171

3) 359 − 295

4) 286 − 214

5) 215 − 197

6) 277 − 181

7) 210 − 145

8) 344 − 222

9) 213 − 164

10) 242 − 156

11) 226 − 152

12) 271 − 172

13) 322 − 173

14) 340 − 306

15) 306 − 263

16) 289 − 217

17) 225 − 213

18) 333 − 253

19) 165 − 164

20) 187 − 131

21) 190 − 170

22) 254 − 151

23) 248 − 161

24) 309 − 290

25) 245 − 146

26) 200 − 193

27) 285 − 267

28) 268 − 226

29) 222 − 133

30) 261 − 122

1)
```
  423
- 285
------
```

2)
```
  388
- 301
------
```

3)
```
  386
- 291
------
```

4)
```
  386
- 313
------
```

5)
```
  406
- 394
------
```

6)
```
  422
- 236
------
```

7)
```
  346
- 277
------
```

8)
```
  345
- 328
------
```

9)
```
  359
- 245
------
```

10)
```
  384
- 291
------
```

11)
```
  446
- 417
------
```

12)
```
  386
- 373
------
```

13)
```
  376
- 360
------
```

14)
```
  433
- 255
------
```

15)
```
  379
- 342
------
```

16)
```
  290
- 241
------
```

17)
```
  384
- 309
------
```

18)
```
  315
- 300
------
```

19)
```
  419
- 268
------
```

20)
```
  401
- 269
------
```

21)
```
  313
- 247
------
```

22)
```
  426
- 320
------
```

23)
```
  389
- 265
------
```

24)
```
  403
- 254
------
```

25)
```
  446
- 422
------
```

26)
```
  408
- 340
------
```

27)
```
  417
- 317
------
```

28)
```
  286
- 232
------
```

29)
```
  435
- 330
------
```

30)
```
  448
- 429
------
```

1) 403
  - 375
  [____]

2) 388
  - 311
  [____]

3) 305
  - 292
  [____]

4) 387
  - 312
  [____]

5) 412
  - 242
  [____]

6) 364
  - 244
  [____]

7) 341
  - 296
  [____]

8) 367
  - 231
  [____]

9) 291
  - 290
  [____]

10) 402
  - 380
  [____]

11) 389
  - 226
  [____]

12) 312
  - 277
  [____]

13) 310
  - 295
  [____]

14) 421
  - 266
  [____]

15) 313
  - 259
  [____]

16) 357
  - 346
  [____]

17) 252
  - 252
  [____]

18) 324
  - 263
  [____]

19) 384
  - 298
  [____]

20) 417
  - 239
  [____]

21) 353
  - 323
  [____]

22) 451
  - 304
  [____]

23) 381
  - 361
  [____]

24) 383
  - 331
  [____]

25) 340
  - 302
  [____]

26) 362
  - 334
  [____]

27) 449
  - 330
  [____]

28) 301
  - 264
  [____]

29) 389
  - 268
  [____]

30) 405
  - 371
  [____]

1) 418
 - 250

2) 497
 - 309

3) 441
 - 270

4) 366
 - 338

5) 432
 - 408

6) 462
 - 452

7) 468
 - 269

8) 387
 - 280

9) 419
 - 323

10) 390
 - 279

11) 464
 - 282

12) 449
 - 417

13) 455
 - 306

14) 406
 - 319

15) 462
 - 401

16) 374
 - 272

17) 382
 - 382

18) 464
 - 421

19) 491
 - 252

20) 487
 - 436

21) 361
 - 281

22) 480
 - 388

23) 462
 - 411

24) 498
 - 417

25) 396
 - 337

26) 386
 - 287

27) 365
 - 336

28) 371
 - 285

29) 434
 - 251

30) 406
 - 354

1) 534
  - 430

2) 522
  - 502

3) 491
  - 416

4) 481
  - 380

5) 525
  - 428

6) 402
  - 401

7) 469
  - 443

8) 481
  - 470

9) 479
  - 460

10) 498
   - 419

11) 541
   - 498

12) 413
   - 401

13) 539
   - 519

14) 464
   - 349

15) 501
   - 454

16) 428
   - 345

17) 502
   - 427

18) 498
   - 455

19) 524
   - 393

20) 396
   - 376

21) 545
   - 470

22) 402
   - 381

23) 484
   - 399

24) 469
   - 355

25) 481
   - 405

26) 479
   - 470

27) 525
   - 492

28) 460
   - 427

29) 523
   - 418

30) 515
   - 430

1)
$$651 - 650$$

2)
$$546 - 410$$

3)
$$400 - 386$$

4)
$$594 - 526$$

5)
$$576 - 564$$

6)
$$539 - 503$$

7)
$$569 - 373$$

8)
$$592 - 569$$

9)
$$579 - 405$$

10)
$$488 - 436$$

11)
$$599 - 590$$

12)
$$583 - 469$$

13)
$$669 - 574$$

14)
$$632 - 411$$

15)
$$546 - 439$$

16)
$$655 - 525$$

17)
$$614 - 528$$

18)
$$632 - 407$$

19)
$$466 - 463$$

20)
$$556 - 443$$

21)
$$468 - 353$$

22)
$$487 - 361$$

23)
$$532 - 486$$

24)
$$648 - 413$$

25)
$$609 - 375$$

26)
$$664 - 420$$

27)
$$484 - 460$$

28)
$$503 - 500$$

29)
$$513 - 350$$

30)
$$527 - 515$$

1) 634 - 509

2) 505 - 491

3) 540 - 400

4) 676 - 562

5) 762 - 570

6) 751 - 558

7) 572 - 435

8) 564 - 526

9) 735 - 697

10) 691 - 659

11) 624 - 429

12) 622 - 573

13) 546 - 488

14) 636 - 491

15) 724 - 460

16) 727 - 422

17) 576 - 515

18) 721 - 495

19) 585 - 428

20) 653 - 514

21) 742 - 632

22) 745 - 464

23) 622 - 610

24) 624 - 572

25) 560 523

26) 748 - 735

27) 634 - 549

28) 688 - 566

29) 584 - 505

30) 535 - 463

1) 628
  - 609

2) 616
  - 603

3) 736
  - 530

4) 869
  - 717

5) 691
  - 512

6) 824
  - 488

7) 786
  - 734

8) 817
  - 675

9) 789
  - 690

10) 768
  - 509

11) 786
  - 598

12) 543
  - 528

13) 823
  - 676

14) 855
  - 549

15) 672
  - 621

16) 832
  - 574

17) 575
  - 514

18) 702
  - 678

19) 866
  - 656

20) 837
  - 731

21) 832
  - 622

22) 686
  - 637

23) 830
  - 487

24) 757
  - 497

25) 614
  - 565

26) 821
  - 577

27) 763
  - 549

28) 819
  - 774

29) 842
  - 713

30) 629
  - 506

NAME:---------DATE:---------

TIME:--------- SCORE: /30

1)
```
  701
- 588
```

2)
```
  918
- 790
```

3)
```
  980
- 609
```

4)
```
  835
- 678
```

5)
```
  647
- 557
```

6)
```
  728
- 511
```

7)
```
  719
- 712
```

8)
```
  982
- 978
```

9)
```
  968
- 714
```

10)
```
  802
- 721
```

11)
```
  839
- 807
```

12)
```
  975
- 550
```

13)
```
  927
- 795
```

14)
```
  913
- 645
```

15)
```
  930
- 597
```

16)
```
  834
- 496
```

17)
```
  936
- 544
```

18)
```
  851
- 503
```

19)
```
  864
- 659
```

20)
```
  809
- 666
```

21)
```
  848
- 667
```

22)
```
  959
  876
```

23)
```
  978
- 630
```

24)
```
  955
- 723
```

25)
```
  897
- 879
```

26)
```
  702
- 493
```

27)
```
  682
- 561
```

28)
```
  974
- 554
```

29)
```
  745
- 543
```

30)
```
  760
- 551
```

Multiplication

1) $8 \times 6 =$ 2) $4 \times 2 =$ 3) $2 \times 6 =$

4) $7 \times 9 =$ 5) $3 \times 5 =$ 6) $6 \times 4 =$

7) $7 \times 4 =$ 8) $7 \times 2 =$ 9) $6 \times 3 =$

10) $7 \times 6 =$ 11) $5 \times 2 =$ 12) $7 \times 3 =$

13) $3 \times 4 =$ 14) $6 \times 2 =$ 15) $7 \times 8 =$

16) $6 \times 8 =$ 17) $2 \times 6 =$ 18) $6 \times 7 =$

19) $7 \times 7 =$ 20) $9 \times 6 =$ 21) $3 \times 4 =$

22) $3 \times 8 =$ 23) $6 \times 4 =$ 24) $8 \times 9 =$

25) $5 \times 9 =$ 26) $7 \times 6 =$ 27) $4 \times 9 =$

28) $9 \times 7 =$ 29) $7 \times 7 =$ 30) $8 \times 7 =$

1) $3 \times 7 =$ 2) $7 \times 8 =$ 3) $7 \times 4 =$

4) $6 \times 2 =$ 5) $6 \times 8 =$ 6) $6 \times 6 =$

7) $9 \times 9 =$ 8) $9 \times 2 =$ 9) $6 \times 7 =$

10) $9 \times 2 =$ 11) $7 \times 6 =$ 12) $3 \times 2 =$

13) $3 \times 5 =$ 14) $4 \times 2 =$ 15) $6 \times 8 =$

16) $6 \times 2 =$ 17) $9 \times 3 =$ 18) $7 \times 7 =$

19) $4 \times 8 =$ 20) $3 \times 8 =$ 21) $3 \times 7 =$

22) $3 \times 4 =$ 23) $7 \times 3 =$ 24) $4 \times 8 =$

25) $8 \times 8 =$ 26) $7 \times 8 =$ 27) $3 \times 4 =$

28) $4 \times 7 =$ 29) $7 \times 4 =$ 30) $8 \times 8 =$

1) $2 \times 9 =$ 2) $4 \times 5 =$ 3) $6 \times 5 =$

4) $6 \times 9 =$ 5) $8 \times 9 =$ 6) $2 \times 5 =$

7) $9 \times 8 =$ 8) $4 \times 7 =$ 9) $2 \times 6 =$

10) $4 \times 5 =$ 11) $9 \times 7 =$ 12) $6 \times 6 =$

13) $2 \times 6 =$ 14) $5 \times 6 =$ 15) $7 \times 7 =$

16) $2 \times 5 =$ 17) $5 \times 8 =$ 18) $8 \times 7 =$

19) $6 \times 8 =$ 20) $9 \times 9 =$ 21) $4 \times 8 =$

22) $5 \times 9 =$ 23) $2 \times 5 =$ 24) $5 \times 6 =$

25) $7 \times 8 =$ 26) $2 \times 8 =$ 27) $5 \times 8 =$

28) $9 \times 7 =$ 29) $9 \times 8 =$ 30) $7 \times 8 =$

NAME:--------- DATE:---------

TIME:--------- SCORE: /30

1) $12 \times 6 =$

2) $15 \times 7 =$

3) $11 \times 3 =$

4) $5 \times 2 =$

5) $10 \times 7 =$

6) $10 \times 2 =$

7) $15 \times 6 =$

8) $10 \times 4 =$

9) $13 \times 5 =$

10) $3 \times 2 =$

11) $5 \times 6 =$

12) $2 \times 6 =$

13) $2 \times 2 =$

14) $4 \times 8 =$

15) $15 \times 5 =$

16) $9 \times 8 =$

17) $13 \times 2 =$

18) $14 \times 7 =$

19) $4 \times 2 =$

20) $13 \times 3 =$

21) $10 \times 4 =$

22) $12 \times 5 =$

23) $8 \times 4 =$

24) $13 \times 2 =$

25) $7 \times 8 =$

26) $15 \times 4 =$

27) $9 \times 4 =$

28) $3 \times 3 =$

29) $7 \times 2 =$

30) $4 \times 8 =$

1) $20 \times 8 =$

2) $9 \times 9 =$

3) $10 \times 8 =$

4) $15 \times 7 =$

5) $7 \times 9 =$

6) $19 \times 4 =$

7) $4 \times 9 =$

8) $5 \times 9 =$

9) $5 \times 3 =$

10) $19 \times 2 =$

11) $13 \times 7 =$

12) $15 \times 9 =$

13) $5 \times 3 =$

14) $7 \times 3 =$

15) $4 \times 8 =$

16) $19 \times 4 =$

17) $13 \times 5 =$

18) $20 \times 7 =$

19) $15 \times 5 =$

20) $6 \times 6 =$

21) $11 \times 2 =$

22) $4 \times 9 =$

23) $14 \times 9 =$

24) $12 \times 9 =$

25) $19 \times 9 =$

26) $5 \times 6 =$

27) $7 \times 3 =$

28) $6 \times 6 =$

29) $17 \times 7 =$

30) $9 \times 6 =$

1) 28 × 5

2) 25 × 3

3) 13 × 6

4) 7 × 8

5) 27 × 7

6) 12 × 8

7) 29 × 5

8) 29 × 3

9) 13 × 7

10) 21 × 9

11) 9 × 3

12) 13 × 6

13) 19 × 7

14) 26 × 9

15) 10 × 4

16) 19 × 5

17) 20 × 3

18) 21 × 3

19) 9 × 4

20) 26 × 7

21) 29 × 8

22) 27 × 8

23) 29 × 3

24) 10 × 4

25) 18 × 3

26) 11 × 6

27) 18 × 8

28) 29 × 3

29) 28 × 9

30) 10 × 7

1) 49
  × 6

2) 70
  × 7

3) 42
  × 5

4) 14
  × 7

5) 26
  × 9

6) 10
  × 3

7) 26
  × 3

8) 18
  × 5

9) 54
  × 3

10) 24
  × 6

11) 36
  × 8

12) 37
  × 4

13) 15
  × 6

14) 42
  × 7

15) 49
  × 6

16) 26
  × 3

17) 49
  × 4

18) 70
  × 5

19) 41
  × 4

20) 61
  × 5

21) 46
  × 6

22) 31
  × 6

23) 37
  × 5

24) 48
  × 6

25) 19
  × 3

26) 27
  × 8

27) 26
  × 6

28) 25
  × 5

29) 62
  × 4

30) 37
  × 7

1) 43
  × 8

2) 52
  × 4

3)   9
  × 3

4) 74
  × 9

5) 66
  × 3

6) 36
  × 8

7) 39
  × 5

8) 72
  × 4

9) 11
  × 8

10) 44
  × 6

11) 12
  × 6

12) 25
  × 6

13) 37
  × 9

14) 48
  × 9

15) 18
  × 8

16) 73
  × 6

17) 78
  × 6

18) 21
  × 9

19) 63
  × 5

20) 18
  × 3

21) 53
  × 4

22) 68
  × 8

23) 57
  × 3

24) 80
  × 7

25) 55
  × 7

26) 52
  × 6

27) 37
  × 4

28) 48
  × 9

29) 58
  × 6

30) 28
  × 8

1) 83
  × 9

2) 60
  × 2

3) 58
  × 2

4) 89
  × 8

5) 30
  × 4

6) 43
  × 6

7) 36
  × 6

8) 49
  × 3

9) 15
  × 2

10) 70
   × 3

11) 33
   × 5

12) 61
   × 4

13) 82
   × 2

14) 90
   × 7

15) 83
   × 7

16) 59
   × 3

17) 19
   × 4

18) 65
   × 3

19) 33
   × 9

20) 45
   × 4

21) 17
   × 3

22) 87
   × 8

23) 54
   × 4

24) 35
   × 5

25) 70
   × 8

26) 42
   × 9

27) 44
   × 7

28) 34
   × 8

29) 41
   × 2

30) 77
   × 2

NAME:--------DATE:--------

DAY:50   TIME:-------- SCORE: /30

1)	90 × 8	2)	82 × 7	3)	49 × 7	4)	76 × 5	5)	47 × 5
6)	31 × 7	7)	90 × 5	8)	69 × 6	9)	60 × 8	10)	88 × 9
11)	46 × 8	12)	82 × 5	13)	57 × 9	14)	28 × 6	15)	34 × 7
16)	80 × 6	17)	40 × 7	18)	85 × 5	19)	49 × 8	20)	98 × 7
21)	64 × 7	22)	67 × 5	23)	55 × 8	24)	89 × 9	25)	87 × 7
26)	84 × 9	27)	72 × 6	28)	85 × 5	29)	32 × 7	30)	61 × 7

1)
$$150 \times 9$$

2)
$$93 \times 9$$

3)
$$50 \times 4$$

4)
$$70 \times 6$$

5)
$$139 \times 7$$

6)
$$104 \times 2$$

7)
$$77 \times 3$$

8)
$$64 \times 3$$

9)
$$54 \times 6$$

10)
$$143 \times 7$$

11)
$$152 \times 6$$

12)
$$121 \times 8$$

13)
$$41 \times 9$$

14)
$$104 \times 5$$

15)
$$98 \times 9$$

16)
$$85 \times 4$$

17)
$$21 \times 5$$

18)
$$104 \times 4$$

19)
$$88 \times 6$$

20)
$$99 \times 7$$

21)
$$58 \times 5$$

22)
$$120 \times 6$$

23)
$$151 \times 8$$

24)
$$69 \times 6$$

25)
$$131 \times 5$$

26)
$$33 \times 3$$

27)
$$150 \times 5$$

28)
$$46 \times 7$$

29)
$$128 \times 4$$

30)
$$124 \times 2$$

31)
$$154 \times 5$$

32)
$$160 \times 7$$

33)
$$133 \times 6$$

34)
$$89 \times 2$$

35)
$$94 \times 4$$

36)
$$144 \times 8$$

1) 202 × 8

2) 137 × 3

3) 186 × 8

4) 64 × 5

5) 187 × 5

6) 190 × 8

7) 203 × 6

8) 183 × 9

9) 165 × 8

10) 147 × 6

11) 91 × 4

12) 69 × 5

13) 80 × 2

14) 228 × 7

15) 124 × 8

16) 51 × 9

17) 249 × 8

18) 240 × 5

19) 193 × 4

20) 150 × 3

21) 102 × 3

22) 153 × 3

23) 237 × 4

24) 141 × 6

25) 165 × 9

26) 146 × 3

27) 233 × 6

28) 124 × 4

29) 74 × 4

30) 184 × 4

31) 221 × 7

32) 127 × 4

33) 112 × 9

34) 54 × 7

35) 118 × 6

36) 189 × 3

1) 310
    ×   5
    [   ]

2) 240
    ×   4
    [   ]

3) 320
    ×   3
    [   ]

4) 156
    ×   3
    [   ]

5) 164
    ×   8
    [   ]

6) 346
    ×   4
    [   ]

7) 116
    ×   2
    [   ]

8) 250
    ×   8
    [   ]

9) 233
    ×   7
    [   ]

10) 169
    ×   3
    [   ]

11) 139
    ×   9
    [   ]

12) 176
    ×   4
    [   ]

13) 234
    ×   6
    [   ]

14) 337
    ×   2
    [   ]

15) 292
    ×   6
    [   ]

16) 325
    ×   3
    [   ]

17) 148
    ×   7
    [   ]

18) 272
    ×   8
    [   ]

19) 285
    ×   8
    [   ]

20) 128
    ×   6
    [   ]

21) 252
    ×   3
    [   ]

22) 339
    ×   8
    [   ]

23) 318
    ×   2
    [   ]

24) 290
    ×   3
    [   ]

25) 346
    ×   8
    [   ]

26) 238
    ×   7
    [   ]

27) 209
    ×   3
    [   ]

28) 110
    ×   9
    [   ]

29) 104
    ×   7
    [   ]

30) 275
    ×   4
    [   ]

31) 223
    ×   8
    [   ]

32) 206
    ×   5
    [   ]

33) 238
    ×   7
    [   ]

34) 237
    ×   5
    [   ]

35) 158
    ×   7
    [   ]

36) 359
    ×   4
    [   ]

1)
```
   480
 ×   9
┌──────┐
└──────┘
```

2)
```
   596
 ×   7
┌──────┐
└──────┘
```

3)
```
   431
 ×   4
┌──────┐
└──────┘
```

4)
```
   213
 ×   6
┌──────┐
└──────┘
```

5)
```
   235
 ×   9
┌──────┐
└──────┘
```

6)
```
   485
 ×   4
┌──────┐
└──────┘
```

7)
```
   357
 ×   7
┌──────┐
└──────┘
```

8)
```
   430
 ×   6
┌──────┐
└──────┘
```

9)
```
   435
 ×   5
┌──────┐
└──────┘
```

10)
```
   462
 ×   3
┌──────┐
└──────┘
```

11)
```
   535
 ×   9
┌──────┐
└──────┘
```

12)
```
   398
 ×   2
┌──────┐
└──────┘
```

13)
```
   508
 ×   3
┌──────┐
└──────┘
```

14)
```
   373
 ×   5
┌──────┐
└──────┘
```

15)
```
   273
 ×   7
┌──────┐
└──────┘
```

16)
```
   535
 ×   5
┌──────┐
└──────┘
```

17)
```
   404
 ×   5
┌──────┐
└──────┘
```

18)
```
   455
 ×   4
┌──────┐
└──────┘
```

19)
```
   210
 ×   9
┌──────┐
└──────┘
```

20)
```
   215
 ×   8
┌──────┐
└──────┘
```

21)
```
   583
 ×   6
┌──────┐
└──────┘
```

22)
```
   461
 ×   8
┌──────┐
└──────┘
```

23)
```
   371
 ×   9
┌──────┐
└──────┘
```

24)
```
   226
 ×   7
┌──────┐
└──────┘
```

25)
```
   357
 ×   2
┌──────┐
└──────┘
```

26)
```
   457
 ×   3
┌──────┐
└──────┘
```

27)
```
   382
 ×   6
┌──────┐
└──────┘
```

28)
```
   284
 ×   4
┌──────┐
└──────┘
```

29)
```
   403
 ×   2
┌──────┐
└──────┘
```

30)
```
   552
 ×   9
┌──────┐
└──────┘
```

31)
```
   293
 ×   5
┌──────┐
└──────┘
```

32)
```
   227
 ×   6
┌──────┐
└──────┘
```

33)
```
   500
 ×   4
┌──────┐
└──────┘
```

34)
```
   500
 ×   9
┌──────┐
└──────┘
```

35)
```
   325
 ×   5
┌──────┐
└──────┘
```

36)
```
   454
 ×   3
┌──────┐
└──────┘
```

1) $\begin{array}{r} 14 \\ \times\ 9 \\ \hline \end{array}$
2) $\begin{array}{r} 29 \\ \times\ 10 \\ \hline \end{array}$
3) $\begin{array}{r} 30 \\ \times\ 8 \\ \hline \end{array}$
4) $\begin{array}{r} 34 \\ \times\ 10 \\ \hline \end{array}$
5) $\begin{array}{r} 23 \\ \times\ 11 \\ \hline \end{array}$
6) $\begin{array}{r} 19 \\ \times\ 12 \\ \hline \end{array}$

7) $\begin{array}{r} 33 \\ \times\ 8 \\ \hline \end{array}$
8) $\begin{array}{r} 35 \\ \times\ 10 \\ \hline \end{array}$
9) $\begin{array}{r} 32 \\ \times\ 12 \\ \hline \end{array}$
10) $\begin{array}{r} 45 \\ \times\ 8 \\ \hline \end{array}$
11) $\begin{array}{r} 25 \\ \times\ 11 \\ \hline \end{array}$
12) $\begin{array}{r} 25 \\ \times\ 9 \\ \hline \end{array}$

13) $\begin{array}{r} 32 \\ \times\ 8 \\ \hline \end{array}$
14) $\begin{array}{r} 10 \\ \times\ 8 \\ \hline \end{array}$
15) $\begin{array}{r} 27 \\ \times\ 12 \\ \hline \end{array}$
16) $\begin{array}{r} 18 \\ \times\ 11 \\ \hline \end{array}$
17) $\begin{array}{r} 25 \\ \times\ 10 \\ \hline \end{array}$
18) $\begin{array}{r} 18 \\ \times\ 11 \\ \hline \end{array}$

19) $\begin{array}{r} 19 \\ \times\ 12 \\ \hline \end{array}$
20) $\begin{array}{r} 32 \\ \times\ 11 \\ \hline \end{array}$
21) $\begin{array}{r} 13 \\ \times\ 9 \\ \hline \end{array}$
22) $\begin{array}{r} 10 \\ \times\ 11 \\ \hline \end{array}$
23) $\begin{array}{r} 17 \\ \times\ 10 \\ \hline \end{array}$
24) $\begin{array}{r} 24 \\ \times\ 12 \\ \hline \end{array}$

25) $\begin{array}{r} 32 \\ \times\ 9 \\ \hline \end{array}$
26) $\begin{array}{r} 11 \\ \times\ 11 \\ \hline \end{array}$
27) $\begin{array}{r} 36 \\ \times\ 10 \\ \hline \end{array}$
28) $\begin{array}{r} 27 \\ \times\ 12 \\ \hline \end{array}$
29) $\begin{array}{r} 37 \\ \times\ 11 \\ \hline \end{array}$
30) $\begin{array}{r} 34 \\ \times\ 12 \\ \hline \end{array}$

31) $\begin{array}{r} 35 \\ \times\ 11 \\ \hline \end{array}$
32) $\begin{array}{r} 22 \\ \times\ 10 \\ \hline \end{array}$
33) $\begin{array}{r} 38 \\ \times\ 8 \\ \hline \end{array}$
34) $\begin{array}{r} 49 \\ \times\ 11 \\ \hline \end{array}$
35) $\begin{array}{r} 33 \\ \times\ 9 \\ \hline \end{array}$
36) $\begin{array}{r} 11 \\ \times\ 8 \\ \hline \end{array}$

1) 24
  × 9

2) 27
  × 11

3) 41
  × 10

4) 60
  × 12

5) 67
  × 12

6) 26
  × 9

7) 54
  × 11

8) 47
  × 9

9) 63
  × 9

10) 32
   × 10

11) 37
   × 9

12) 55
   × 12

13) 53
   × 12

14) 48
   × 8

15) 34
   × 12

16) 29
   × 11

17) 46
   × 10

18) 48
   × 9

19) 69
   × 12

20) 39
   × 10

21) 26
   × 9

22) 54
   × 9

23) 58
   × 11

24) 55
   × 11

25) 31
   × 10

26) 69
   × 9

27) 22
   × 11

28) 42
   × 8

29) 49
   × 11

30) 53
   × 12

1) 48
  × 10

2) 95
  × 12

3) 85
  × 11

4) 30
  × 10

5) 31
  × 11

6) 44
  × 10

7) 47
  × 12

8) 69
  × 11

9) 22
  × 10

10) 96
  × 11

11) 93
  × 12

12) 34
  × 10

13) 34
  × 12

14) 57
  × 11

15) 97
  × 12

16) 43
  × 11

17) 73
  × 11

18) 69
  × 12

19) 73
  × 11

20) 30
  × 12

21) 42
  × 11

22) 23
  × 10

23) 36
  × 10

24) 95
  × 10

25) 60
  × 12

26) 71
  × 11

27) 34
  × 12

28) 37
  × 12

29) 97
  × 11

30) 64
  × 11

1) 156
  × 12

2) 134
  × 11

3)   65
  ×   9

4) 115
  × 11

5) 151
  × 11

6) 191
  × 11

7)   96
  × 10

8) 189
  × 10

9) 126
  × 11

10) 157
  ×   9

11) 183
  ×   8

12) 92
  ×   9

13) 152
  ×   8

14) 189
  ×   9

15) 158
  × 10

16) 84
  ×   9

17) 176
  ×   8

18) 149
  × 12

19) 160
  × 10

20) 111
  ×   8

21) 64
  × 10

22) 56
  ×   8

23) 119
  × 10

24) 186
  ×   8

25) 135
  ×   8

26) 183
  × 12

27) 88
  × 11

28) 146
  ×   9

29) 178
  × 11

30) 130
  × 12

1) 175
  × 10

2) 168
  × 12

3) 389
  × 11

4) 352
  × 10

5) 374
  × 10

6) 139
  × 10

7) 112
  × 12

8) 211
  × 11

9) 387
  × 11

10) 169
  × 11

11) 337
  × 12

12) 249
  × 12

13) 166
  × 9

14) 157
  × 12

15) 247
  × 9

16) 253
  × 10

17) 298
  × 9

18) 159
  × 9

19) 191
  × 12

20) 188
  × 12

21) 205
  × 9

22) 367
  × 9

23) 103
  × 9

24) 299
  × 9

25) 354
  × 9

26) 267
  × 12

27) 115
  × 10

28) 295
  × 9

29) 283
  × 12

30) 198
  × 9

1) 480
  × 10

2) 658
  × 10

3) 499
  × 10

4) 451
  × 11

5) 732
  × 10

6) 705
  × 12

7) 573
  × 10

8) 499
  × 12

9) 687
  × 10

10) 532
  × 12

11) 669
  × 10

12) 678
  × 12

13) 717
  × 11

14) 785
  × 12

15) 466
  × 11

16) 649
  × 12

17) 742
  × 10

18) 691
  × 11

19) 530
  × 12

20) 731
  × 11

21) 708
  × 11

22) 717
  × 12

23) 478
  × 11

24) 492
  × 12

25) 573
  × 12

26) 550
  × 11

27) 497
  × 12

28) 464
  × 12

29) 502
  × 12

30) 628
  × 10

Division

NAME:--------- DATE:---------

TIME:--------- SCORE: /30

1) $5 \overline{)2\ 5}$

2) $8 \overline{)5\ 6}$

3) $2 \overline{)6}$

4) $8 \overline{)4\ 8}$

5) $8 \overline{)3\ 2}$

6) $2 \overline{)3\ 8}$

7) $6 \overline{)4\ 2}$

8) $2 \overline{)1\ 4}$

9) $2 \overline{)2}$

10) $4 \overline{)1\ 6}$

11) $2 \overline{)1\ 0}$

12) $7 \overline{)1\ 4}$

13) $3 \overline{)4\ 8}$

14) $5 \overline{)2\ 5}$

15) $8 \overline{)8}$

16) $6 \overline{)5\ 4}$

17) $5 \overline{)1\ 5}$

18) $6 \overline{)3\ 6}$

19) $6 \overline{)6}$

20) $5 \overline{)5\ 5}$

21) $5 \overline{)4\ 0}$

22) $3 \overline{)1\ 5}$

23) $3 \overline{)3\ 3}$

24) $2 \overline{)2\ 2}$

25) $9 \overline{)4\ 5}$

26) $3 \overline{)1\ 2}$

27) $2 \overline{)2\ 8}$

28) $2 \overline{)2\ 0}$

29) $2 \overline{)2}$

30) $6 \overline{)1\ 2}$

1) $3 \overline{)5\ 1}$

2) $4 \overline{)8\ 0}$

3) $4 \overline{)4\ 0}$

4) $4 \overline{)1\ 2}$

5) $4 \overline{)5\ 6}$

6) $8 \overline{)6\ 4}$

7) $3 \overline{)5\ 7}$

8) $8 \overline{)8}$

9) $3 \overline{)5\ 7}$

10) $3 \overline{)9}$

11) $4 \overline{)6\ 8}$

12) $3 \overline{)9}$

13) $4 \overline{)5\ 2}$

14) $3 \overline{)3\ 9}$

15) $7 \overline{)1\ 4}$

16) $3 \overline{)9}$

17) $8 \overline{)2\ 4}$

18) $3 \overline{)4\ 2}$

19) $7 \overline{)3\ 5}$

20) $5 \overline{)7\ 5}$

21) $4 \overline{)7\ 6}$

22) $8 \overline{)8\ 0}$

23) $3 \overline{)4\ 5}$

24) $6 \overline{)7\ 8}$

25) $3 \overline{)7\ 8}$

26) $4 \overline{)8}$

27) $4 \overline{)1\ 6}$

28) $7 \overline{)7\ 7}$

29) $5 \overline{)5\ 5}$

30) $3 \overline{)1\ 2}$

NAME:--------DATE:---------

TIME:-------- SCORE: /30

1) $5 \overline{)8\ 0}$ 2) $4 \overline{)7\ 6}$ 3) $3 \overline{)3\ 9}$ 4) $5 \overline{)8\ 0}$ 5) $5 \overline{)6\ 5}$

6) $4 \overline{)3\ 2}$ 7) $3 \overline{)8\ 7}$ 8) $3 \overline{)5\ 7}$ 9) $8 \overline{)5\ 6}$ 10) $3 \overline{)4\ 5}$

11) $3 \overline{)3\ 0}$ 12) $9 \overline{)9\ 9}$ 13) $8 \overline{)7\ 2}$ 14) $5 \overline{)3\ 5}$ 15) $5 \overline{)7\ 0}$

16) $3 \overline{)3\ 0}$ 17) $4 \overline{)4\ 4}$ 18) $3 \overline{)1\ 5}$ 19) $7 \overline{)7\ 7}$ 20) $3 \overline{)6\ 9}$

21) $3 \overline{)3\ 0}$ 22) $5 \overline{)5\ 5}$ 23) $6 \overline{)4\ 8}$ 24) $8 \overline{)9\ 6}$ 25) $7 \overline{)4\ 9}$

26) $3 \overline{)6\ 9}$ 27) $3 \overline{)5\ 1}$ 28) $4 \overline{)7\ 2}$ 29) $4 \overline{)6\ 0}$ 30) $4 \overline{)1\ 2}$

1) $6\overline{)8\ 4}$ 2) $5\overline{)4\ 5}$ 3) $8\overline{)2\ 4}$ 4) $7\overline{)4\ 9}$ 5) $3\overline{)9\ 9}$

6) $5\overline{)9\ 0}$ 7) $4\overline{)5\ 2}$ 8) $5\overline{)6\ 0}$ 9) $9\overline{)9\ 0}$ 10) $5\overline{)8\ 5}$

11) $4\overline{)9\ 2}$ 12) $3\overline{)5\ 7}$ 13) $5\overline{)7\ 0}$ 14) $4\overline{)9\ 2}$ 15) $3\overline{)9\ 9}$

16) $5\overline{)2\ 5}$ 17) $3\overline{)5\ 7}$ 18) $5\overline{)6\ 5}$ 19) $6\overline{)3\ 6}$ 20) $8\overline{)3\ 2}$

21) $3\overline{)5\ 7}$ 22) $7\overline{)9\ 1}$ 23) $7\overline{)9\ 1}$ 24) $7\overline{)7\ 7}$ 25) $3\overline{)7\ 5}$

26) $5\overline{)5\ 5}$ 27) $5\overline{)8\ 5}$ 28) $8\overline{)4\ 0}$ 29) $4\overline{)4\ 4}$ 30) $4\overline{)9\ 6}$

1) 8)3 2   2) 5)7 5   3) 9)3 6   4) 5)4 5   5) 6)8 4

6) 8)7 2   7) 7)2 1   8) 9)8 1   9) 7)1 4   10) 8)1 6

11) 5)7 0   12) 5)6 5   13) 6)7 8   14) 9)5 4   15) 5)3 0

16) 7)3 5   17) 9)8 1   18) 5)2 5   19) 5)8 5   20) 5)4 0

21) 7)4 9   22) 8)4 8   23) 5)9 0   24) 5)3 5   25) 9)9 9

26) 8)4 8   27) 7)9 1   28) 7)2 8   29) 6)8 4   30) 6)7 8

1) $3\overline{)84}$ 2) $3\overline{)87}$ 3) $5\overline{)55}$ 4) $5\overline{)35}$ 5) $3\overline{)45}$

6) $3\overline{)75}$ 7) $5\overline{)70}$ 8) $5\overline{)75}$ 9) $6\overline{)96}$ 10) $5\overline{)55}$

11) $5\overline{)10}$ 12) $8\overline{)96}$ 13) $6\overline{)18}$ 14) $7\overline{)98}$ 15) $9\overline{)81}$

16) $4\overline{)68}$ 17) $4\overline{)16}$ 18) $6\overline{)12}$ 19) $6\overline{)96}$ 20) $3\overline{)42}$

21) $3\overline{)54}$ 22) $6\overline{)12}$ 23) $3\overline{)87}$ 24) $5\overline{)20}$ 25) $4\overline{)48}$

26) $5\overline{)10}$ 27) $8\overline{)80}$ 28) $8\overline{)88}$ 29) $6\overline{)30}$ 30) $6\overline{)24}$

1) $6\overline{)144}$ 2) $3\overline{)93}$ 3) $3\overline{)33}$ 4) $3\overline{)51}$

5) $5\overline{)75}$ 6) $5\overline{)150}$ 7) $6\overline{)30}$ 8) $4\overline{)116}$

9) $7\overline{)84}$ 10) $4\overline{)16}$ 11) $8\overline{)120}$ 12) $3\overline{)69}$

13) $3\overline{)78}$ 14) $5\overline{)70}$ 15) $6\overline{)78}$ 16) $3\overline{)57}$

17) $9\overline{)27}$ 18) $6\overline{)120}$ 19) $4\overline{)96}$ 20) $5\overline{)85}$

21) $4\overline{)92}$ 22) $3\overline{)39}$ 23) $4\overline{)92}$ 24) $5\overline{)130}$

1) 6)7 2

2) 8)6 4

3) 4)1 5 6

4) 4)1 1 6

5) 7)4 9

6) 5)1 2 5

7) 5)1 1 0

8) 7)3 5

9) 9)1 3 5

10) 4)5 2

11) 3)1 8 6

12) 5)5 0

13) 3)8 7

14) 6)1 3 8

15) 4)5 2

16) 4)1 4 8

17) 4)4 4

18) 8)6 4

19) 3)4 2

20) 8)8 8

21) 6)1 6 2

22) 3)9 6

23) 4)5 2

24) 4)1 5 2

1) 6)3 0 0   2) 4)1 7 2   3) 7)3 8 5   4) 3)2 9 1

5) 8)2 9 6   6) 4)3 9 2   7) 7)2 1 7   8) 3)3 9 9

9) 7)2 3 8   10) 8)1 2 8   11) 7)2 6 6   12) 5)1 6 5

13) 7)3 7 1   14) 7)3 2 2   15) 3)9 0   16) 9)3 2 4

17) 3)3 1 2   18) 8)1 1 2   19) 3)2 3 1   20) 7)2 1 7

21) 7)3 5 0   22) 4)1 6 4   23) 5)1 4 0   24) 3)8 7

1) 2)3 6 6

2) 4)2 7 2

3) 5)1 7 5

4) 5)3 7 5

5) 5)3 9 5

6) 7)1 3 3

7) 2)2 0 0

8) 5)2 9 5

9) 7)2 1 7

10) 2)2 4 8

11) 9)1 8 0

12) 3)4 8 6

13) 2)2 9 4

14) 2)3 0 8

15) 9)3 5 1

16) 2)1 3 0

17) 2)2 2 6

18) 3)2 7 9

19) 3)2 6 4

20) 8)4 4 8

21) 7)4 6 2

22) 6)1 6 2

23) 6)3 7 8

24) 7)1 8 9

1) $3 \overline{)4\ 6\ 5}$ 2) $3 \overline{)2\ 8\ 5}$ 3) $5 \overline{)2\ 7\ 5}$ 4) $7 \overline{)3\ 2\ 2}$

5) $7 \overline{)4\ 7\ 6}$ 6) $3 \overline{)4\ 5}$ 7) $3 \overline{)4\ 2\ 9}$ 8) $7 \overline{)1\ 6\ 1}$

9) $5 \overline{)1\ 2\ 5}$ 10) $5 \overline{)3\ 4\ 5}$ 11) $3 \overline{)1\ 8\ 3}$ 12) $6 \overline{)1\ 0\ 8}$

13) $3 \overline{)3\ 6\ 0}$ 14) $3 \overline{)2\ 0\ 1}$ 15) $3 \overline{)4\ 6\ 8}$ 16) $6 \overline{)4\ 7\ 4}$

17) $3 \overline{)4\ 2\ 6}$ 18) $3 \overline{)3\ 6\ 3}$ 19) $4 \overline{)3\ 8\ 8}$ 20) $5 \overline{)1\ 1\ 5}$

21) $4 \overline{)4\ 2\ 8}$ 22) $3 \overline{)2\ 0\ 1}$ 23) $7 \overline{)4\ 6\ 9}$ 24) $3 \overline{)9\ 9}$

1) $8 \overline{)1\ 1\ 2}$

2) $7 \overline{)1\ 8\ 2}$

3) $4 \overline{)3\ 3\ 6}$

4) $8 \overline{)4\ 3\ 2}$

5) $4 \overline{)1\ 6}$

6) $4 \overline{)4\ 3\ 6}$

7) $8 \overline{)8\ 8}$

8) $3 \overline{)4\ 7\ 1}$

9) $4 \overline{)3\ 6\ 4}$

10) $8 \overline{)2\ 0\ 8}$

11) $5 \overline{)4\ 8\ 5}$

12) $3 \overline{)3\ 9\ 6}$

13) $5 \overline{)1\ 8\ 5}$

14) $3 \overline{)3\ 1\ 8}$

15) $4 \overline{)7\ 6}$

16) $7 \overline{)2\ 4\ 5}$

17) $3 \overline{)2\ 4\ 0}$

18) $5 \overline{)1\ 3\ 0}$

19) $3 \overline{)2\ 2\ 2}$

20) $7 \overline{)2\ 6\ 6}$

21) $9 \overline{)2\ 4\ 3}$

22) $3 \overline{)3\ 9\ 3}$

23) $7 \overline{)4\ 0\ 6}$

24) $5 \overline{)1\ 6\ 5}$

1) $3 \overline{)72}$

2) $9 \overline{)81}$

3) $5 \overline{)55}$

4) $12 \overline{)36}$

5) $9 \overline{)63}$

6) $5 \overline{)20}$

7) $9 \overline{)18}$

8) $10 \overline{)20}$

9) $7 \overline{)49}$

10) $7 \overline{)14}$

11) $7 \overline{)63}$

12) $4 \overline{)68}$

13) $3 \overline{)15}$

14) $7 \overline{)91}$

15) $4 \overline{)16}$

16) $8 \overline{)16}$

17) $4 \overline{)12}$

18) $7 \overline{)63}$

19) $10 \overline{)50}$

20) $7 \overline{)21}$

21) $3 \overline{)33}$

22) $10 \overline{)30}$

23) $9 \overline{)99}$

24) $3 \overline{)87}$

1) $4\overline{)68}$ 2) $3\overline{)21}$ 3) $3\overline{)42}$ 4) $9\overline{)90}$

5) $3\overline{)96}$ 6) $9\overline{)72}$ 7) $7\overline{)49}$ 8) $11\overline{)22}$

9) $5\overline{)35}$ 10) $3\overline{)93}$ 11) $4\overline{)12}$ 12) $3\overline{)27}$

13) $6\overline{)36}$ 14) $5\overline{)85}$ 15) $3\overline{)93}$ 16) $3\overline{)33}$

17) $6\overline{)66}$ 18) $3\overline{)48}$ 19) $9\overline{)54}$ 20) $11\overline{)99}$

21) $8\overline{)88}$ 22) $3\overline{)84}$ 23) $3\overline{)69}$ 24) $3\overline{)57}$

1) 8)6 4 0   2) 8)2 1 6   3) 6)4 3 8   4) 9)4 7 7

5) 7)6 7 9   6) 9)1 7 1   7) 8)6 8 8   8) 8)2 7 2

9) 7)6 5 8   10) 7)5 5 3   11) 8)4 7 2   12) 5)5 8 0

13) 8)5 1 2   14) 7)1 4 7   15) 8)2 5 6   16) 9)3 3 3

17) 5)6 1 5   18) 5)6 0 5   19) 9)6 2 1   20) 9)5 1 3

21) 7)6 5 1   22) 5)5 0 0   23) 9)6 3 9   24) 7)1 3 3

1) 3)5 1 9
2) 4)5 2
3) 6)3 1 8
4) 4)1 9 6

5) 8)4 7 2
6) 6)3 4 2
7) 4)2 3 6
8) 3)1 1 7

9) 5)3 3 0
10) 3)6 9
11) 7)4 2 0
12) 9)2 4 3

13) 6)2 1 0
14) 7)6 4 4
15) 6)5 3 4
16) 4)1 8 8

17) 5)1 8 5
18) 3)1 9 8
19) 8)5 3 6
20) 4)2 7 2

21) 3)5 1
22) 3)1 7 4
23) 4)6 5 6
24) 7)6 8 6

86

1) $7 \overline{)4\ 9}$

2) $5 \overline{)9\ 5}$

3) $5 \overline{)6\ 5}$

4) $5 \overline{)9\ 0}$

5) $9 \overline{)4\ 5}$

6) $6 \overline{)9\ 6}$

7) $6 \overline{)7\ 8}$

8) $5 \overline{)9\ 5}$

9) $1\ 1 \overline{)8\ 8}$

10) $5 \overline{)4\ 5}$

11) $5 \overline{)8\ 5}$

12) $7 \overline{)9\ 8}$

13) $5 \overline{)9\ 5}$

14) $5 \overline{)2\ 5}$

15) $1\ 1 \overline{)1\ 1}$

16) $6 \overline{)8\ 4}$

17) $1\ 0 \overline{)8\ 0}$

18) $9 \overline{)6\ 3}$

19) $6 \overline{)5\ 4}$

20) $8 \overline{)4\ 0}$

21) $8 \overline{)4\ 0}$

22) $5 \overline{)9\ 5}$

23) $6 \overline{)1\ 2}$

24) $8 \overline{)1\ 6}$

1)
$$9\overline{)5\ 2\ 2}$$

2)
$$3\overline{)6\ 3\ 9}$$

3)
$$3\overline{)5\ 0\ 1}$$

4)
$$4\overline{)7\ 4\ 8}$$

5)
$$4\overline{)3\ 5\ 2}$$

6)
$$3\overline{)6\ 9\ 0}$$

7)
$$3\overline{)6\ 3\ 9}$$

8)
$$5\overline{)3\ 3\ 5}$$

9)
$$5\overline{)7\ 1\ 5}$$

10)
$$5\overline{)5\ 9\ 5}$$

11)
$$3\overline{)5\ 1\ 9}$$

12)
$$3\overline{)3\ 6\ 6}$$

13)
$$9\overline{)4\ 2\ 3}$$

14)
$$5\overline{)4\ 5\ 5}$$

15)
$$3\overline{)5\ 0\ 1}$$

16)
$$6\overline{)4\ 6\ 2}$$

17)
$$8\overline{)5\ 2\ 0}$$

18)
$$5\overline{)5\ 4\ 5}$$

19)
$$4\overline{)6\ 9\ 2}$$

20)
$$9\overline{)4\ 0\ 5}$$

21)
$$3\overline{)3\ 9\ 3}$$

22)
$$3\overline{)4\ 0\ 8}$$

23)
$$8\overline{)4\ 7\ 2}$$

24)
$$5\overline{)5\ 7\ 5}$$

1) $3\overline{)9\ 2\ 1}$ 2) $7\overline{)8\ 8\ 2}$ 3) $5\overline{)5\ 0\ 0}$ 4) $4\overline{)8\ 8\ 4}$

5) $8\overline{)6\ 0\ 8}$ 6) $7\overline{)4\ 3\ 4}$ 7) $5\overline{)7\ 5\ 5}$ 8) $5\overline{)6\ 1\ 5}$

9) $4\overline{)4\ 1\ 6}$ 10) $4\overline{)8\ 8\ 0}$ 11) $7\overline{)5\ 1\ 1}$ 12) $7\overline{)8\ 6\ 8}$

13) $3\overline{)9\ 0\ 6}$ 14) $7\overline{)9\ 3\ 1}$ 15) $3\overline{)9\ 8\ 4}$ 16) $3\overline{)4\ 1\ 7}$

17) $4\overline{)6\ 2\ 8}$ 18) $8\overline{)5\ 6\ 8}$ 19) $5\overline{)7\ 0\ 0}$ 20) $3\overline{)9\ 0\ 6}$

21) $5\overline{)4\ 4\ 5}$ 22) $7\overline{)6\ 0\ 9}$ 23) $6\overline{)4\ 6\ 2}$ 24) $5\overline{)8\ 3\ 0}$

1)

$$3\overline{)651}$$

2)

$$8\overline{)680}$$

3)

$$4\overline{)836}$$

4)

$$4\overline{)628}$$

5)

$$7\overline{)987}$$

6)

$$4\overline{)680}$$

7)

$$8\overline{)832}$$

8)

$$7\overline{)868}$$

9)

$$3\overline{)897}$$

10)

$$8\overline{)920}$$

11)

$$5\overline{)445}$$

12)

$$3\overline{)606}$$

13)

$$9\overline{)846}$$

14)

$$9\overline{)621}$$

15)

$$8\overline{)760}$$

16)

$$9\overline{)981}$$

17)

$$9\overline{)702}$$

18)

$$6\overline{)564}$$

19)

$$8\overline{)896}$$

20)

$$3\overline{)753}$$

21)

$$3\overline{)534}$$

22)

$$5\overline{)685}$$

23)

$$9\overline{)540}$$

24)

$$7\overline{)861}$$

Fraction

NAME:--------- DATE:---------

TIME:--------- SCORE: /30

1) $\dfrac{9}{8} + \dfrac{12}{11} =$

2) $\dfrac{11}{3} + \dfrac{11}{11} =$

3) $\dfrac{7}{4} + \dfrac{3}{3} =$

4) $\dfrac{8}{8} + \dfrac{11}{10} =$

5) $\dfrac{13}{6} + \dfrac{12}{8} =$

6) $\dfrac{10}{8} + \dfrac{4}{3} =$

7) $\dfrac{15}{14} + \dfrac{2}{2} =$

8) $\dfrac{13}{12} + \dfrac{12}{10} =$

9) $\dfrac{14}{8} + \dfrac{12}{6} =$

10) $\dfrac{11}{6} + \dfrac{12}{7} =$

11) $\dfrac{11}{10} + \dfrac{12}{12} =$

12) $\dfrac{12}{6} + \dfrac{6}{5} =$

13) $\dfrac{5}{4} + \dfrac{9}{6} =$

14) $\dfrac{14}{11} + \dfrac{9}{3} =$

15) $\dfrac{10}{10} + \dfrac{11}{2} =$

16) $\dfrac{13}{5} + \dfrac{10}{7} =$

17) $\dfrac{11}{4} + \dfrac{12}{7} =$

18) $\dfrac{15}{15} + \dfrac{11}{5} =$

19) $\dfrac{15}{15} + \dfrac{12}{9} =$

20) $\dfrac{13}{9} + \dfrac{8}{8} =$

21) $\dfrac{7}{5} + \dfrac{12}{12} =$

22) $\dfrac{12}{8} + \dfrac{8}{4} =$

23) $\dfrac{15}{14} + \dfrac{11}{3} =$

24) $\dfrac{14}{6} + \dfrac{8}{5} =$

25) $\dfrac{15}{2} + \dfrac{11}{7} =$

26) $\dfrac{15}{12} + \dfrac{11}{11} =$

27) $\dfrac{15}{13} + \dfrac{12}{9} =$

28) $\dfrac{13}{12} + \dfrac{12}{10} =$

29) $\dfrac{14}{13} + \dfrac{8}{8} =$

30) $\dfrac{14}{14} + \dfrac{6}{5} =$

1) $\dfrac{15}{12} + \dfrac{12}{8} =$

2) $\dfrac{13}{3} + \dfrac{7}{7} =$

3) $\dfrac{8}{7} + \dfrac{6}{3} =$

4) $\dfrac{6}{2} + \dfrac{12}{8} =$

5) $\dfrac{7}{3} + \dfrac{6}{2} =$

6) $\dfrac{15}{6} + \dfrac{12}{7} =$

7) $\dfrac{14}{13} + \dfrac{12}{10} =$

8) $\dfrac{4}{2} + \dfrac{9}{6} =$

9) $\dfrac{13}{3} + \dfrac{7}{6} =$

10) $\dfrac{10}{9} + \dfrac{5}{4} =$

11) $\dfrac{15}{12} + \dfrac{8}{4} =$

12) $\dfrac{14}{12} + \dfrac{11}{11} =$

13) $\dfrac{10}{4} + \dfrac{12}{12} =$

14) $\dfrac{14}{12} + \dfrac{5}{2} =$

15) $\dfrac{15}{14} + \dfrac{2}{2} =$

16) $\dfrac{15}{10} + \dfrac{7}{7} =$

17) $\dfrac{8}{5} + \dfrac{12}{9} =$

18) $\dfrac{13}{7} + \dfrac{11}{9} =$

19) $\dfrac{5}{3} + \dfrac{12}{5} =$

20) $\dfrac{13}{7} + \dfrac{11}{3} =$

21) $\dfrac{5}{2} + \dfrac{12}{11} =$

22) $\dfrac{14}{8} + \dfrac{10}{9} =$

23) $\dfrac{8}{5} + \dfrac{6}{6} =$

24) $\dfrac{10}{6} + \dfrac{12}{4} =$

25) $\dfrac{8}{6} + \dfrac{8}{4} =$

26) $\dfrac{6}{2} + \dfrac{12}{5} =$

27) $\dfrac{15}{11} + \dfrac{5}{4} =$

28) $\dfrac{9}{6} + \dfrac{12}{10} =$

29) $\dfrac{15}{15} + \dfrac{10}{3} =$

30) $\dfrac{15}{9} + \dfrac{11}{7} =$

1) $\dfrac{13}{4} + \dfrac{11}{10} =$

2) $\dfrac{14}{13} + \dfrac{8}{7} =$

3) $\dfrac{14}{7} + \dfrac{11}{6} =$

4) $\dfrac{7}{5} + \dfrac{11}{3} =$

5) $\dfrac{12}{11} + \dfrac{11}{10} =$

6) $\dfrac{12}{12} + \dfrac{9}{5} =$

7) $\dfrac{12}{5} + \dfrac{12}{12} =$

8) $\dfrac{12}{8} + \dfrac{11}{7} =$

9) $\dfrac{14}{4} + \dfrac{12}{9} =$

10) $\dfrac{14}{6} + \dfrac{11}{10} =$

11) $\dfrac{9}{7} + \dfrac{10}{3} =$

12) $\dfrac{13}{11} + \dfrac{11}{5} =$

13) $\dfrac{15}{15} + \dfrac{10}{7} =$

14) $\dfrac{15}{14} + \dfrac{10}{6} =$

15) $\dfrac{13}{7} + \dfrac{12}{6} =$

16) $\dfrac{12}{10} + \dfrac{5}{3} =$

17) $\dfrac{11}{11} + \dfrac{11}{10} =$

18) $\dfrac{15}{10} + \dfrac{12}{7} =$

19) $\dfrac{8}{6} + \dfrac{9}{9} =$

20) $\dfrac{13}{13} + \dfrac{12}{11} =$

21) $\dfrac{4}{4} + \dfrac{9}{7} =$

22) $\dfrac{11}{5} + \dfrac{12}{11} =$

23) $\dfrac{14}{10} + \dfrac{8}{3} =$

24) $\dfrac{14}{10} + \dfrac{12}{2} =$

25) $\dfrac{10}{3} + \dfrac{2}{2} =$

26) $\dfrac{15}{9} + \dfrac{4}{2} =$

27) $\dfrac{15}{9} + \dfrac{11}{5} =$

28) $\dfrac{14}{14} + \dfrac{10}{7} =$

29) $\dfrac{9}{9} + \dfrac{11}{4} =$

30) $\dfrac{7}{4} + \dfrac{12}{12} =$

1) $1\frac{8}{12} + 1\frac{6}{9} =$

2) $1\frac{5}{15} + 1\frac{3}{6} =$

3) $2\frac{2}{6} + 3\frac{3}{4} =$

4) $1\frac{1}{8} + 1\frac{4}{13} =$

5) $1\frac{5}{13} + 1\frac{3}{8} =$

6) $1\frac{9}{10} + 4\frac{1}{3} =$

7) $1\frac{1}{14} + 1\frac{1}{8} =$

8) $1\frac{8}{10} + 4\frac{1}{3} =$

9) $2\frac{3}{7} + 1\frac{4}{11} =$

10) $8\frac{1}{2} + 1\frac{6}{9} =$

11) $1\frac{3}{6} + 1\frac{1}{4} =$

12) $1\frac{2}{12} + 1\frac{1}{3} =$

13) $2\frac{5}{6} + 1\frac{3}{9} =$

14) $2\frac{3}{8} + 1\frac{9}{11} =$

15) $2\frac{1}{9} + 1\frac{5}{13} =$

16) $1\frac{3}{9} + 1\frac{4}{13} =$

17) $2\frac{3}{8} + 2\frac{2}{9} =$

18) $1\frac{7}{10} + 2\frac{1}{3} =$

19) $1\frac{3}{10} + 3\frac{1}{6} =$

20) $2\frac{2}{8} + 1\frac{6}{9} =$

1) $\dfrac{4}{7} - \dfrac{8}{16} =$

2) $\dfrac{9}{10} - \dfrac{3}{4} =$

3) $\dfrac{6}{14} - \dfrac{1}{17} =$

4) $\dfrac{3}{6} - \dfrac{7}{16} =$

5) $\dfrac{10}{13} - \dfrac{6}{12} =$

6) $\dfrac{3}{6} - \dfrac{5}{17} =$

7) $\dfrac{2}{6} - \dfrac{2}{16} =$

8) $\dfrac{1}{6} - \dfrac{1}{18} =$

9) $\dfrac{2}{5} - \dfrac{3}{14} =$

10) $\dfrac{3}{6} - \dfrac{1}{15} =$

11) $\dfrac{10}{13} - \dfrac{9}{18} =$

12) $\dfrac{7}{20} - \dfrac{1}{6} =$

13) $\dfrac{1}{3} - \dfrac{2}{7} =$

14) $\dfrac{6}{8} - \dfrac{9}{17} =$

15) $\dfrac{1}{2} - \dfrac{7}{15} =$

16) $\dfrac{5}{13} - \dfrac{1}{12} =$

17) $\dfrac{2}{4} - \dfrac{1}{16} =$

18) $\dfrac{1}{2} - \dfrac{2}{20} =$

19) $\dfrac{10}{13} - \dfrac{2}{3} =$

20) $\dfrac{4}{6} - \dfrac{7}{12} =$

21) $\dfrac{7}{9} - \dfrac{1}{4} =$

22) $\dfrac{6}{10} - \dfrac{4}{18} =$

23) $\dfrac{2}{13} - \dfrac{1}{11} =$

24) $\dfrac{2}{3} - \dfrac{2}{16} =$

25) $\dfrac{6}{16} - \dfrac{2}{10} =$

26) $\dfrac{4}{8} - \dfrac{3}{18} =$

27) $\dfrac{8}{11} - \dfrac{1}{2} =$

28) $\dfrac{10}{11} - \dfrac{5}{17} =$

29) $\dfrac{9}{14} - \dfrac{2}{16} =$

30) $\dfrac{2}{15} - \dfrac{1}{16} =$

NAME:--------DATE:--------

TIME:-------- SCORE: /30

1) $\dfrac{3}{19} - \dfrac{1}{20} =$

2) $\dfrac{8}{20} - \dfrac{2}{10} =$

3) $\dfrac{7}{15} - \dfrac{8}{19} =$

4) $\dfrac{1}{2} - \dfrac{1}{3} =$

5) $\dfrac{3}{5} - \dfrac{9}{18} =$

6) $\dfrac{6}{10} - \dfrac{5}{13} =$

7) $\dfrac{1}{8} - \dfrac{1}{19} =$

8) $\dfrac{6}{10} - \dfrac{4}{15} =$

9) $\dfrac{10}{18} - \dfrac{8}{15} =$

10) $\dfrac{2}{3} - \dfrac{6}{13} =$

11) $\dfrac{5}{13} - \dfrac{3}{19} =$

12) $\dfrac{3}{12} - \dfrac{1}{5} =$

13) $\dfrac{2}{4} - \dfrac{6}{19} =$

14) $\dfrac{1}{2} - \dfrac{2}{16} =$

15) $\dfrac{9}{17} - \dfrac{1}{8} =$

16) $\dfrac{1}{3} - \dfrac{3}{13} =$

17) $\dfrac{7}{14} - \dfrac{2}{10} =$

18) $\dfrac{2}{8} - \dfrac{3}{14} =$

19) $\dfrac{3}{4} - \dfrac{2}{9} =$

20) $\dfrac{6}{13} - \dfrac{5}{19} =$

21) $\dfrac{7}{8} - \dfrac{2}{3} =$

22) $\dfrac{9}{17} - \dfrac{1}{2} =$

23) $\dfrac{7}{19} - \dfrac{1}{18} =$

24) $\dfrac{2}{3} - \dfrac{6}{15} =$

25) $\dfrac{3}{5} - \dfrac{3}{7} =$

26) $\dfrac{2}{14} - \dfrac{1}{11} =$

27) $\dfrac{7}{9} - \dfrac{2}{10} =$

28) $\dfrac{8}{9} - \dfrac{5}{8} =$

29) $\dfrac{2}{12} - \dfrac{1}{9} =$

30) $\dfrac{8}{19} - \dfrac{3}{11} =$

1) $\dfrac{5}{9} - \dfrac{7}{13} =$

2) $\dfrac{8}{15} - \dfrac{1}{11} =$

3) $\dfrac{10}{13} - \dfrac{6}{19} =$

4) $\dfrac{9}{13} - \dfrac{6}{16} =$

5) $\dfrac{9}{10} - \dfrac{6}{19} =$

6) $\dfrac{2}{16} - \dfrac{2}{19} =$

7) $\dfrac{2}{3} - \dfrac{1}{18} =$

8) $\dfrac{2}{3} - \dfrac{2}{13} =$

9) $\dfrac{9}{10} - \dfrac{2}{16} =$

10) $\dfrac{2}{4} - \dfrac{2}{17} =$

11) $\dfrac{8}{12} - \dfrac{1}{2} =$

12) $\dfrac{4}{8} - \dfrac{4}{11} =$

13) $\dfrac{7}{8} - \dfrac{2}{5} =$

14) $\dfrac{5}{6} - \dfrac{2}{3} =$

15) $\dfrac{3}{5} - \dfrac{5}{11} =$

16) $\dfrac{5}{15} - \dfrac{2}{10} =$

17) $\dfrac{8}{20} - \dfrac{6}{19} =$

18) $\dfrac{3}{4} - \dfrac{2}{5} =$

19) $\dfrac{4}{6} - \dfrac{1}{2} =$

20) $\dfrac{2}{3} - \dfrac{2}{11} =$

21) $\dfrac{7}{8} - \dfrac{7}{12} =$

22) $\dfrac{7}{16} - \dfrac{3}{9} =$

23) $\dfrac{6}{20} - \dfrac{1}{5} =$

24) $\dfrac{6}{7} - \dfrac{5}{9} -$

25) $\dfrac{2}{4} - \dfrac{5}{19} =$

26) $\dfrac{8}{12} - \dfrac{1}{4} =$

27) $\dfrac{9}{17} - \dfrac{10}{20} =$

28) $\dfrac{7}{15} - \dfrac{4}{16} =$

29) $\dfrac{8}{19} - \dfrac{1}{5} =$

30) $\dfrac{1}{2} - \dfrac{7}{15} =$

1) $1 \frac{6}{11} - 1 \frac{2}{14} =$

2) $1 \frac{4}{5} - 1 \frac{4}{13} =$

3) $1 \frac{3}{12} - 1 \frac{3}{14} =$

4) $4 \frac{3}{4} - 1 \frac{2}{14} =$

5) $7 \frac{1}{2} - 1 \frac{3}{15} =$

6) $1 \frac{5}{12} - 1 \frac{3}{9} =$

7) $2 \frac{2}{8} - 1 \frac{1}{14} =$

8) $2 \frac{2}{3} - 1 \frac{9}{10} =$

9) $3 \frac{1}{2} - 1 \frac{2}{9} =$

10) $1 \frac{6}{7} - 1 \frac{2}{12} =$

11) $1 \frac{1}{3} - 1 \frac{1}{10} =$

12) $3 \frac{2}{6} - 1 \frac{2}{15} =$

13) $3 \frac{2}{6} - 1 \frac{5}{11} =$

14) $2 \frac{4}{6} - 2 \frac{1}{2} =$

15) $5 \frac{2}{3} - 4 \frac{3}{4} =$

16) $6 \frac{1}{2} - 1 \frac{2}{12} =$

17) $1 \frac{8}{11} - 1 \frac{3}{8} =$

18) $1 \frac{4}{6} - 1 \frac{7}{12} =$

19) $1 \frac{6}{7} - 1 \frac{8}{12} =$

20) $1 \frac{9}{11} - 1 \frac{4}{10} =$

1) $\dfrac{5}{14} \times \dfrac{6}{11} =$

2) $\dfrac{1}{13} \times \dfrac{1}{12} =$

3) $\dfrac{1}{7} \times \dfrac{5}{12} =$

4) $\dfrac{4}{14} \times \dfrac{1}{12} =$

5) $\dfrac{8}{10} \times \dfrac{4}{5} =$

6) $\dfrac{2}{9} \times \dfrac{2}{4} =$

7) $\dfrac{2}{6} \times \dfrac{6}{12} =$

8) $\dfrac{2}{4} \times \dfrac{1}{2} =$

9) $\dfrac{3}{6} \times \dfrac{3}{8} =$

10) $\dfrac{2}{5} \times \dfrac{1}{4} =$

11) $\dfrac{2}{15} \times \dfrac{1}{2} =$

12) $\dfrac{8}{14} \times \dfrac{10}{12} =$

13) $\dfrac{7}{9} \times \dfrac{3}{4} =$

14) $\dfrac{2}{10} \times \dfrac{3}{8} =$

15) $\dfrac{1}{2} \times \dfrac{8}{11} =$

16) $\dfrac{8}{11} \times \dfrac{4}{5} =$

17) $\dfrac{10}{13} \times \dfrac{5}{12} =$

18) $\dfrac{2}{12} \times \dfrac{5}{7} =$

19) $\dfrac{2}{11} \times \dfrac{3}{7} =$

20) $\dfrac{2}{9} \times \dfrac{5}{10} =$

21) $\dfrac{2}{12} \times \dfrac{5}{6} =$

22) $\dfrac{5}{10} \times \dfrac{3}{8} =$

23) $\dfrac{2}{11} \times \dfrac{4}{8} =$

24) $\dfrac{4}{10} \times \dfrac{8}{12} =$

25) $\dfrac{2}{12} \times \dfrac{6}{9} =$

26) $\dfrac{1}{15} \times \dfrac{6}{8} =$

27) $\dfrac{5}{13} \times \dfrac{8}{10} =$

28) $\dfrac{1}{5} \times \dfrac{3}{11} =$

29) $\dfrac{4}{6} \times \dfrac{6}{7} =$

30) $\dfrac{3}{5} \times \dfrac{5}{8} =$

NAME:--------DATE:--------

TIME:--------- SCORE: /30

1) $\dfrac{1}{10} \times \dfrac{5}{9} =$

2) $\dfrac{1}{16} \times \dfrac{3}{4} =$

3) $\dfrac{1}{19} \times \dfrac{7}{12} =$

4) $\dfrac{5}{6} \times \dfrac{10}{20} =$

5) $\dfrac{7}{17} \times \dfrac{4}{12} =$

6) $\dfrac{10}{18} \times \dfrac{4}{6} =$

7) $\dfrac{11}{18} \times \dfrac{2}{9} =$

8) $\dfrac{10}{13} \times \dfrac{10}{20} =$

9) $\dfrac{3}{9} \times \dfrac{12}{13} =$

10) $\dfrac{1}{2} \times \dfrac{4}{8} =$

11) $\dfrac{4}{8} \times \dfrac{2}{3} =$

12) $\dfrac{2}{7} \times \dfrac{13}{20} =$

13) $\dfrac{1}{3} \times \dfrac{8}{12} =$

14) $\dfrac{2}{16} \times \dfrac{6}{19} =$

15) $\dfrac{15}{20} \times \dfrac{3}{17} =$

16) $\dfrac{5}{7} \times \dfrac{3}{11} =$

17) $\dfrac{9}{18} \times \dfrac{3}{6} =$

18) $\dfrac{3}{6} \times \dfrac{7}{20} =$

19) $\dfrac{2}{16} \times \dfrac{2}{8} =$

20) $\dfrac{2}{5} \times \dfrac{9}{18} =$

21) $\dfrac{4}{12} \times \dfrac{12}{15} =$

22) $\dfrac{6}{11} \times \dfrac{7}{15} =$

23) $\dfrac{12}{19} \times \dfrac{8}{18} =$

24) $\dfrac{1}{3} \times \dfrac{10}{12} =$

25) $\dfrac{3}{20} \times \dfrac{1}{2} =$

26) $\dfrac{6}{13} \times \dfrac{4}{5} =$

27) $\dfrac{3}{8} \times \dfrac{1}{2} =$

28) $\dfrac{4}{14} \times \dfrac{1}{2} =$

29) $\dfrac{1}{2} \times \dfrac{4}{12} =$

30) $\dfrac{3}{16} \times \dfrac{14}{20} =$

1) $\dfrac{6}{17} \times \dfrac{1}{3} =$

2) $\dfrac{1}{2} \times \dfrac{12}{17} =$

3) $\dfrac{2}{14} \times \dfrac{14}{15} =$

4) $\dfrac{3}{11} \times \dfrac{7}{17} =$

5) $\dfrac{3}{9} \times \dfrac{7}{8} =$

6) $\dfrac{4}{5} \times \dfrac{8}{14} =$

7) $\dfrac{5}{15} \times \dfrac{7}{17} =$

8) $\dfrac{2}{8} \times \dfrac{3}{4} =$

9) $\dfrac{3}{13} \times \dfrac{2}{3} =$

10) $\dfrac{2}{8} \times \dfrac{2}{9} =$

11) $\dfrac{13}{14} \times \dfrac{7}{8} =$

12) $\dfrac{14}{20} \times \dfrac{11}{18} =$

13) $\dfrac{11}{15} \times \dfrac{6}{20} =$

14) $\dfrac{3}{14} \times \dfrac{7}{20} =$

15) $\dfrac{10}{11} \times \dfrac{2}{15} =$

16) $\dfrac{1}{2} \times \dfrac{3}{13} =$

17) $\dfrac{6}{8} \times \dfrac{6}{17} =$

18) $\dfrac{4}{19} \times \dfrac{2}{6} =$

19) $\dfrac{7}{11} \times \dfrac{14}{18} =$

20) $\dfrac{9}{17} \times \dfrac{4}{12} =$

21) $\dfrac{12}{19} \times \dfrac{3}{14} =$

22) $\dfrac{1}{18} \times \dfrac{2}{3} =$

23) $\dfrac{14}{17} \times \dfrac{9}{11} =$

24) $\dfrac{2}{7} \times \dfrac{1}{2} =$

25) $\dfrac{6}{11} \times \dfrac{4}{13} =$

26) $\dfrac{11}{15} \times \dfrac{1}{12} =$

27) $\dfrac{12}{17} \times \dfrac{11}{14} =$

28) $\dfrac{1}{19} \times \dfrac{13}{14} =$

29) $\dfrac{3}{11} \times \dfrac{4}{19} =$

30) $\dfrac{14}{15} \times \dfrac{2}{19} =$

1) $1\dfrac{5}{15} \times 1\dfrac{4}{10} =$

2) $1\dfrac{2}{8} \times 1\dfrac{6}{11} =$

3) $1\dfrac{8}{12} \times 1\dfrac{1}{15} =$

4) $9\dfrac{1}{2} \times 1\dfrac{7}{10} =$

5) $6\dfrac{1}{2} \times 1\dfrac{1}{9} =$

6) $3\dfrac{1}{5} \times 1\dfrac{5}{12} =$

7) $1\dfrac{2}{9} \times 2\dfrac{2}{5} =$

8) $1\dfrac{6}{9} \times 4\dfrac{1}{2} =$

9) $1\dfrac{1}{14} \times 1\dfrac{9}{11} =$

10) $4\dfrac{1}{3} \times 1\dfrac{2}{6} =$

11) $4\dfrac{1}{3} \times 1\dfrac{3}{15} =$

12) $6\dfrac{1}{2} \times 2\dfrac{1}{5} =$

13) $1\dfrac{7}{9} \times 1\dfrac{8}{12} =$

14) $1\dfrac{1}{13} \times 2\dfrac{1}{7} =$

15) $2\dfrac{1}{6} \times 1\dfrac{4}{8} =$

16) $1\dfrac{1}{14} \times 1\dfrac{3}{13} =$

17) $6\dfrac{1}{3} \times 1\dfrac{2}{9} =$

18) $1\dfrac{5}{11} \times 3\dfrac{1}{6} =$

19) $1\dfrac{3}{13} \times 1\dfrac{3}{14} =$

20) $2\dfrac{1}{5} \times 1\dfrac{5}{9} =$

1) $\dfrac{5}{4} \div \dfrac{5}{5} =$

2) $\dfrac{5}{2} \div \dfrac{8}{5} =$

3) $\dfrac{5}{3} \div \dfrac{9}{6} =$

4) $\dfrac{5}{4} \div \dfrac{10}{2} =$

5) $\dfrac{7}{2} \div \dfrac{11}{8} =$

6) $\dfrac{7}{6} \div \dfrac{8}{3} =$

7) $\dfrac{9}{4} \div \dfrac{9}{6} =$

8) $\dfrac{9}{9} \div \dfrac{12}{4} =$

9) $\dfrac{8}{4} \div \dfrac{10}{8} =$

10) $\dfrac{8}{8} \div \dfrac{9}{6} =$

11) $\dfrac{8}{8} \div \dfrac{11}{2} =$

12) $\dfrac{9}{9} \div \dfrac{7}{5} =$

13) $\dfrac{7}{4} \div \dfrac{11}{11} =$

14) $\dfrac{4}{3} \div \dfrac{11}{11} =$

15) $\dfrac{8}{3} \div \dfrac{10}{10} =$

16) $\dfrac{9}{5} \div \dfrac{11}{10} =$

17) $\dfrac{6}{4} \div \dfrac{7}{2} =$

18) $\dfrac{7}{2} \div \dfrac{9}{9} =$

19) $\dfrac{2}{2} \div \dfrac{6}{3} =$

20) $\dfrac{7}{6} \div \dfrac{8}{4} =$

21) $\dfrac{9}{2} \div \dfrac{12}{6} =$

22) $\dfrac{8}{7} \div \dfrac{12}{4} =$

23) $\dfrac{4}{4} \div \dfrac{9}{7} =$

24) $\dfrac{9}{8} \div \dfrac{4}{3} =$

25) $\dfrac{8}{7} \div \dfrac{9}{8} =$

26) $\dfrac{8}{6} \div \dfrac{7}{7} =$

27) $\dfrac{9}{6} \div \dfrac{12}{10} =$

28) $\dfrac{8}{5} \div \dfrac{6}{6} =$

29) $\dfrac{9}{8} \div \dfrac{11}{4} =$

30) $\dfrac{9}{2} \div \dfrac{8}{7} =$

1) $\dfrac{11}{11} \div \dfrac{10}{3} =$

2) $\dfrac{4}{3} \div \dfrac{13}{10} =$

3) $\dfrac{11}{5} \div \dfrac{7}{2} =$

4) $\dfrac{12}{9} \div \dfrac{13}{13} =$

5) $\dfrac{12}{12} \div \dfrac{7}{5} =$

6) $\dfrac{6}{2} \div \dfrac{11}{4} =$

7) $\dfrac{9}{9} \div \dfrac{14}{11} =$

8) $\dfrac{8}{4} \div \dfrac{7}{3} =$

9) $\dfrac{9}{2} \div \dfrac{15}{4} =$

10) $\dfrac{10}{10} \div \dfrac{7}{3} =$

11) $\dfrac{9}{2} \div \dfrac{11}{3} =$

12) $\dfrac{12}{5} \div \dfrac{13}{13} =$

13) $\dfrac{12}{12} \div \dfrac{13}{6} =$

14) $\dfrac{11}{11} \div \dfrac{13}{8} =$

15) $\dfrac{9}{9} \div \dfrac{10}{4} =$

16) $\dfrac{11}{4} \div \dfrac{10}{2} =$

17) $\dfrac{11}{4} \div \dfrac{12}{8} =$

18) $\dfrac{7}{6} \div \dfrac{15}{7} =$

19) $\dfrac{9}{3} \div \dfrac{12}{11} =$

20) $\dfrac{6}{5} \div \dfrac{8}{7} =$

21) $\dfrac{11}{11} \div \dfrac{15}{14} =$

22) $\dfrac{10}{8} \div \dfrac{6}{4} =$

23) $\dfrac{11}{11} \div \dfrac{5}{2} =$

24) $\dfrac{11}{6} \div \dfrac{14}{12} =$

25) $\dfrac{8}{6} \div \dfrac{14}{14} =$

26) $\dfrac{9}{5} \div \dfrac{10}{9} =$

27) $\dfrac{12}{12} \div \dfrac{9}{5} =$

28) $\dfrac{9}{5} \div \dfrac{14}{14} =$

29) $\dfrac{11}{9} \div \dfrac{8}{3} =$

30) $\dfrac{9}{8} \div \dfrac{12}{6} =$

1) $\dfrac{11}{11} \div \dfrac{11}{9} =$

2) $\dfrac{10}{8} \div \dfrac{14}{9} =$

3) $\dfrac{7}{6} \div \dfrac{15}{14} =$

4) $\dfrac{11}{11} \div \dfrac{14}{12} =$

5) $\dfrac{11}{9} \div \dfrac{5}{5} =$

6) $\dfrac{10}{6} \div \dfrac{15}{5} =$

7) $\dfrac{4}{2} \div \dfrac{13}{8} =$

8) $\dfrac{12}{12} \div \dfrac{13}{10} =$

9) $\dfrac{7}{3} \div \dfrac{12}{8} =$

10) $\dfrac{12}{7} \div \dfrac{15}{3} =$

11) $\dfrac{12}{9} \div \dfrac{15}{15} =$

12) $\dfrac{12}{12} \div \dfrac{14}{7} =$

13) $\dfrac{11}{11} \div \dfrac{14}{9} =$

14) $\dfrac{8}{4} \div \dfrac{15}{8} =$

15) $\dfrac{5}{5} \div \dfrac{14}{7} =$

16) $\dfrac{12}{11} \div \dfrac{4}{3} =$

17) $\dfrac{6}{6} \div \dfrac{10}{8} =$

18) $\dfrac{9}{9} \div \dfrac{9}{8} =$

19) $\dfrac{7}{5} \div \dfrac{12}{7} =$

20) $\dfrac{9}{4} \div \dfrac{15}{15} =$

21) $\dfrac{11}{10} \div \dfrac{14}{14} =$

22) $\dfrac{10}{3} \div \dfrac{12}{12} =$

23) $\dfrac{8}{6} \div \dfrac{15}{14} =$

24) $\dfrac{12}{12} \div \dfrac{7}{4} =$

25) $\dfrac{9}{7} \div \dfrac{13}{13} =$

26) $\dfrac{10}{9} \div \dfrac{7}{4} =$

27) $\dfrac{6}{2} \div \dfrac{13}{12} =$

28) $\dfrac{12}{5} \div \dfrac{14}{14} =$

29) $\dfrac{10}{4} \div \dfrac{5}{5} =$

30) $\dfrac{12}{10} \div \dfrac{13}{12} =$

1) $7\frac{1}{2} \div 1\frac{3}{11} =$

2) $1\frac{2}{8} \div 1\frac{5}{12} =$

3) $1\frac{4}{5} \div 1\frac{5}{11} =$

4) $2\frac{1}{7} \div 5\frac{1}{2} =$

5) $1\frac{3}{15} \div 2\frac{3}{7} =$

6) $1\frac{7}{13} \div 1\frac{3}{4} =$

7) $1\frac{4}{9} \div 2\frac{5}{6} =$

8) $1\frac{1}{7} \div 2\frac{1}{8} =$

9) $1\frac{1}{9} \div 2\frac{2}{6} =$

10) $6\frac{2}{3} \div 2\frac{1}{5} =$

11) $1\frac{6}{8} \div 3\frac{1}{3} =$

12) $2\frac{4}{6} \div 3\frac{2}{3} =$

13) $2\frac{2}{4} \div 1\frac{2}{6} =$

14) $1\frac{3}{13} \div 1\frac{6}{8} =$

15) $4\frac{2}{4} \div 1\frac{3}{5} =$

16) $9\frac{1}{2} \div 1\frac{3}{9} =$

17) $1\frac{4}{15} \div 2\frac{1}{9} =$

18) $3\frac{1}{5} \div 1\frac{3}{4} =$

19) $1\frac{7}{12} \div 1\frac{1}{6} =$

20) $1\frac{3}{13} \div 1\frac{2}{7} =$

1) $\dfrac{\square}{\square}$ =

2) $\dfrac{\square}{\square}$ =

3) $\dfrac{\square}{\square}$ =

4) $\dfrac{\square}{\square}$ =

5) $\dfrac{\square}{\square}$ =

6) $\dfrac{\square}{\square}$ =

7) $\dfrac{\square}{\square}$ =

8) $\dfrac{\square}{\square}$ =

9) $\dfrac{\square}{\square}$ =

10) $\dfrac{\square}{\square}$ =

11) $\dfrac{\square}{\square}$ =

12) $\dfrac{\square}{\square}$ =

13) $\dfrac{\square}{\square}$ =

14) $\dfrac{\square}{\square}$ =

15) $\dfrac{\square}{\square}$ =

16) $\dfrac{\square}{\square}$ =

17) $\dfrac{\square}{\square}$ =

18) $\dfrac{\square}{\square}$ =

19) $\dfrac{\square}{\square}$ =

20) $\dfrac{\square}{\square}$ =

21) $\dfrac{\square}{\square}$ =

22) $\dfrac{\square}{\square}$ =

23) $\dfrac{\square}{\square}$ =

24) $\dfrac{\square}{\square}$ =

25) $\dfrac{\square}{\square}$ =

26) $\dfrac{\square}{\square}$ =

27) $\dfrac{\square}{\square}$ =

28) $\dfrac{\square}{\square}$ =

29) $\dfrac{\square}{\square}$ =

30) $\dfrac{\square}{\square}$ =

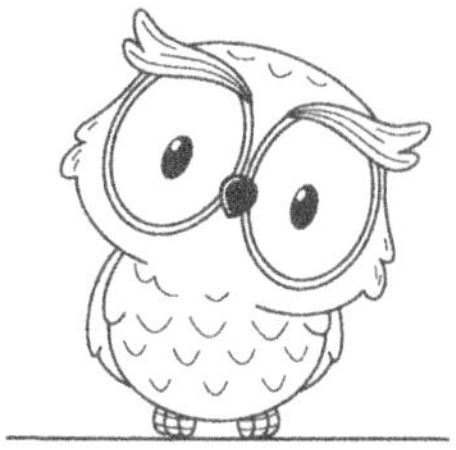

NAME:---------DATE:---------

DAY:98 TIME:--------- SCORE: /30

1) $\frac{\square}{\square}$ =

2) $\frac{\square}{\square}$ =

3) $\frac{\square}{\square}$ =

4) $\frac{\square}{\square}$ =

5) $\frac{\square}{\square}$ =

6) $\frac{\square}{\square}$ =

7) $\frac{\square}{\square}$ =

8) $\frac{\square}{\square}$ =

9) $\frac{\square}{\square}$ =

10) $\frac{\square}{\square}$ =

11) $\frac{\square}{\square}$ =

12) $\frac{\square}{\square}$ =

13) $\frac{\square}{\square}$ =

14) $\frac{\square}{\square}$ =

15) $\frac{\square}{\square}$ =

16) $\frac{\square}{\square}$ =

17) $\frac{\square}{\square}$ =

18) $\frac{\square}{\square}$ =

19) $\frac{\square}{\square}$ =

20) $\frac{\square}{\square}$ =

21) $\frac{\square}{\square}$ =

22) $\frac{\square}{\square}$ =

23) $\frac{\square}{\square}$ =

24) $\frac{\square}{\square}$ =

25) $\frac{\square}{\square}$ =

26) $\frac{\square}{\square}$ =

27) $\frac{\square}{\square}$ =

28) $\frac{\square}{\square}$ =

29) $\frac{\square}{\square}$ =

30) $\frac{\square}{\square}$ =

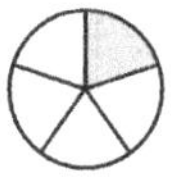

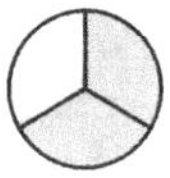

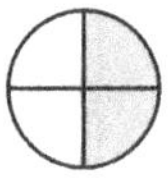

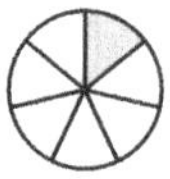

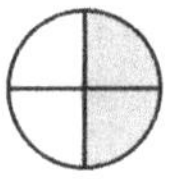

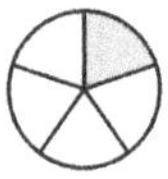

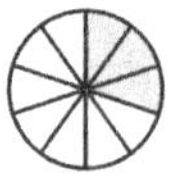

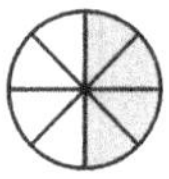

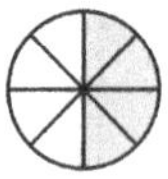

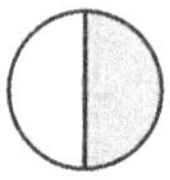

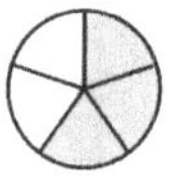

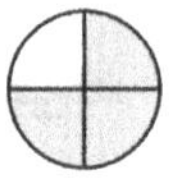

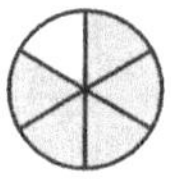

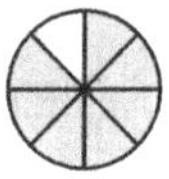

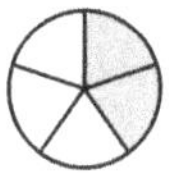

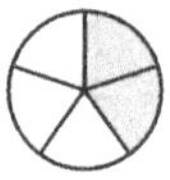

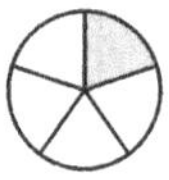

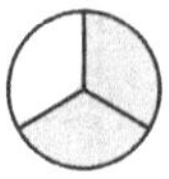

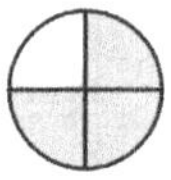

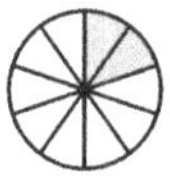

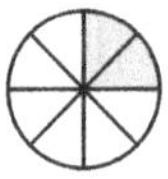

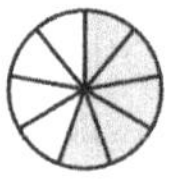

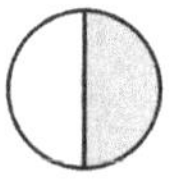

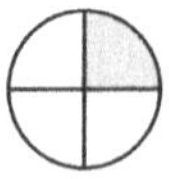

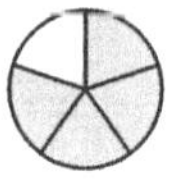

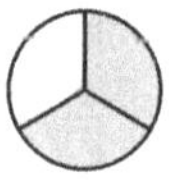

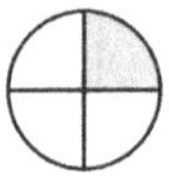

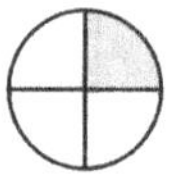

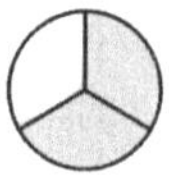

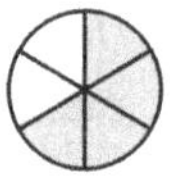

1) $\dfrac{\square}{\square}$ =

2) $\dfrac{\square}{\square}$ =

3) $\dfrac{\square}{\square}$ =

4) $\dfrac{\square}{\square}$ =

5) $\dfrac{\square}{\square}$ =

6) $\dfrac{\square}{\square}$ =

7) $\dfrac{\square}{\square}$ =

8) $\dfrac{\square}{\square}$ =

9) $\dfrac{\square}{\square}$ =

10) $\dfrac{\square}{\square}$ =

11) $\dfrac{\square}{\square}$ =

12) $\dfrac{\square}{\square}$ =

13) $\dfrac{\square}{\square}$ =

14) $\dfrac{\square}{\square}$ =

15) $\dfrac{\square}{\square}$ =

16) $\dfrac{\square}{\square}$ =

17) $\dfrac{\square}{\square}$ =

18) $\dfrac{\square}{\square}$ =

19) $\dfrac{\square}{\square}$ =

20) $\dfrac{\square}{\square}$ =

21) $\dfrac{\square}{\square}$ =

22) $\dfrac{\square}{\square}$ =

23) $\dfrac{\square}{\square}$ =

24) $\dfrac{\square}{\square}$ =

25) $\dfrac{\square}{\square}$ =

26) $\dfrac{\square}{\square}$ =

27) $\dfrac{\square}{\square}$ =

28) $\dfrac{\square}{\square}$ =

29) $\dfrac{\square}{\square}$ =

30) $\dfrac{\square}{\square}$ =

1) ▢/▢ = ⊘

2) ▢/▢ = ⊘

3) ▢/▢ = ⊘

4) ▢/▢ = ⊘

5) ▢/▢ = ⊘

6) ▢/▢ = ⊘

7) ▢/▢ = ⊘

8) ▢/▢ = ⊘

9) ▢/▢ = ⊘

10) ▢/▢ = ⊘

11) ▢/▢ = ⊘

12) ▢/▢ = ⊘

13) ▢/▢ = ⊘

14) ▢/▢ = ⊘

15) ▢/▢ = ⊘

16) ▢/▢ = ⊘

17) ▢/▢ = ⊘

18) ▢/▢ = ⊘

19) ▢/▢ = ⊘

20) ▢/▢ = ⊘

21) ▢/▢ = ⊘

22) ▢/▢ = ⊘

23) ▢/▢ = ⊘

24) ▢/▢ = ⊘

25) ▢/▢ = ⊘

26) ▢/▢ = ⊘

27) ▢/▢ = ⊘

28) ▢/▢ = ⊘

29) ▢/▢ = ⊘

30) ▢/▢ = ⊘

Answer

Page 8, Item 1:
(1)187 (2)163 (3)161 (4)181 (5)170 (6)190
(7)185 (8)167 (9)182 (10)182 (11)183
(12)173 (13)171 (14)192 (15)174 (16)182
(17)177 (18)173 (19)171 (20)172 (21)179
(22)192 (23)183 (24)178 (25)182 (26)174
(27)165 (28)171 (29)178 (30)157

Page 9, Item 1:
(1)185 (2)184 (3)167 (4)169 (5)172 (6)176
(7)177 (8)167 (9)174 (10)185 (11)186
(12)184 (13)178 (14)174 (15)173 (16)193
(17)192 (18)187 (19)174 (20)174 (21)187
(22)180 (23)175 (24)196 (25)166 (26)166
(27)170 (28)175 (29)179 (30)173

Page 10, Item 1:
(1)198 (2)185 (3)191 (4)188 (5)183 (6)198
(7)196 (8)162 (9)173 (10)200 (11)193
(12)178 (13)180 (14)192 (15)166 (16)183
(17)187 (18)191 (19)183 (20)167 (21)192
(22)177 (23)193 (24)191 (25)202 (26)164
(27)190 (28)196 (29)165 (30)182

Page 11, Item 1:
(1)198 (2)207 (3)196 (4)168 (5)179 (6)184
(7)203 (8)193 (9)209 (10)217 (11)182
(12)169 (13)171 (14)180 (15)186 (16)175
(17)206 (18)183 (19)182 (20)193 (21)218
(22)171 (23)210 (24)186 (25)205 (26)185
(27)193 (28)205 (29)172 (30)175

Page 12, Item 1:
(1)202 (2)190 (3)200 (4)236 (5)201 (6)196
(7)242 (8)203 (9)216 (10)229 (11)201
(12)214 (13)234 (14)197 (15)223 (16)216
(17)209 (18)221 (19)219 (20)204 (21)212
(22)218 (23)204 (24)227 (25)226 (26)213
(27)228 (28)194 (29)203 (30)220

Page 13, Item 1:
(1)188 (2)187 (3)239 (4)252 (5)184 (6)248
(7)219 (8)197 (9)192 (10)241 (11)196
(12)246 (13)202 (14)215 (15)213 (16)236
(17)201 (18)202 (19)217 (20)176 (21)206
(22)251 (23)199 (24)188 (25)205 (26)191
(27)248 (28)240 (29)238 (30)258

Page 14, Item 1:
(1)234 (2)205 (3)231 (4)215 (5)216 (6)231
(7)259 (8)192 (9)177 (10)243 (11)247
(12)265 (13)197 (14)210 (15)198 (16)201
(17)217 (18)274 (19)177 (20)175 (21)224
(22)192 (23)199 (24)228 (25)214 (26)249
(27)226 (28)204 (29)220 (30)228

Page 15, Item 1:
(1)346 (2)499 (3)475 (4)394 (5)378 (6)416
(7)375 (8)460 (9)419 (10)409 (11)327
(12)437 (13)312 (14)448 (15)422 (16)335
(17)393 (18)373 (19)380 (20)361 (21)457
(22)395 (23)347 (24)467 (25)436 (26)413
(27)370 (28)399 (29)412 (30)379

Page 16, Item 1:
(1)255 (2)252 (3)232 (4)338 (5)317 (6)232
(7)361 (8)234 (9)355 (10)295 (11)301
(12)269 (13)196 (14)237 (15)270 (16)342
(17)188 (18)270 (19)357 (20)262 (21)319
(22)267 (23)189 (24)263 (25)261 (26)261
(27)237 (28)278 (29)261 (30)251

Page 17, Item 1:
(1)311 (2)231 (3)213 (4)420 (5)311 (6)250
(7)316 (8)275 (9)236 (10)208 (11)312
(12)248 (13)334 (14)219 (15)364 (16)310
(17)259 (18)247 (19)235 (20)373 (21)377
(22)356 (23)283 (24)356 (25)190 (26)320
(27)228 (28)324 (29)263 (30)339

Page 18, Item 1:
(1)322 (2)415 (3)288 (4)328 (5)286 (6)355
(7)288 (8)343 (9)346 (10)258 (11)278
(12)369 (13)242 (14)409 (15)334 (16)312
(17)393 (18)257 (19)297 (20)296 (21)307
(22)297 (23)299 (24)328 (25)361 (26)363
(27)331 (28)333 (29)302 (30)293

Page 19, Item 1:
(1)363 (2)380 (3)430 (4)403 (5)315 (6)429
(7)366 (8)376 (9)324 (10)305 (11)445
(12)460 (13)259 (14)300 (15)399 (16)324
(17)236 (18)408 (19)347 (20)439 (21)314
(22)313 (23)458 (24)398 (25)319 (26)311
(27)353 (28)330 (29)310 (30)440

Page 20, Item 1:
(1)325 (2)300 (3)409 (4)263 (5)355 (6)363
(7)383 (8)392 (9)375 (10)319 (11)335
(12)315 (13)252 (14)311 (15)292 (16)332
(17)288 (18)324 (19)338 (20)297 (21)346
(22)246 (23)308 (24)241 (25)312 (26)265
(27)356 (28)249 (29)309 (30)301

Page 21, Item 1:
(1)306 (2)264 (3)337 (4)337 (5)306 (6)332
(7)340 (8)289 (9)367(10)326 (11)238
(12)341 (13)427 (14)218 (15)297 (16)251
(17)355 (18)370 (19)324 (20)313 (21)342
(22)356 (23)334 (24)378 (25)351 (26)358
(27)302 (28)296 (29)337 (30)305

Page 22, Item 1:
(1)635 (2)519 (3)590 (4)603 (5)614 (6)551
(7)725 (8)564 (9)641 (10)602 (11)528
(12)555 (13)625 (14)598 (15)636 (16)626
(17)596 (18)571 (19)555 (20)520 (21)536
(22)618 (23)518 (24)469 (25)508 (26)488
(27)419 (28)519 (29)590 (30)608

Page 23, Item 1:
(1)189 (2)165 (3)231 (4)222 (5)206 (6)178
(7)182 (8)180 (9)180 (10)205 (11)227
(12)199 (13)165 (14)206 (15)190 (16)172
(17)208 (18)173 (19)233 (20)215 (21)198
(22)239 (23)243 (24)162 (25)205 (26)183
(27)174 (28)189 (29)219 (30)168

Page 24, Item 1:
(1)229 (2)214 (3)178 (4)171 (5)183 (6)241
(7)208 (8)202 (9)221 (10)227 (11)186
(12)184 (13)206 (14)230 (15)182 (16)196
(17)218 (18)203 (19)205 (20)167 (21)205
(22)182 (23)211 (24)189 (25)218 (26)220
(27)215 (28)192 (29)187 (30)243

Page 25, Item 1:
(1)227 (2)172 (3)172 (4)231 (5)170 (6)213
(7)178 (8)218 (9)199 (10)194 (11)203
(12)210 (13)168 (14)188 (15)198 (16)176
(17)225 (18)223 (19)218 (20)218 (21)210
(22)178 (23)225 (24)215 (25)199 (26)205
(27)194 (28)199 (29)198 (30)197

Page 26, Item 1:
(1)262 (2)343 (3)187 (4)271 (5)373 (6)262
(7)242 (8)352 (9)356 (10)281 (11)311
(12)232 (13)228 (14)385 (15)352 (16)370
(17)265 (18)331 (19)308 (20)350 (21)280
(22)272 (23)314 (24)251 (25)240 (26)233
(27)370 (28)251 (29)363 (30)328

Page 27, Item 1:
(1)345 (2)334 (3)456 (4)418 (5)255 (6)354
(7)345 (8)437 (9)424 (10)435 (11)326
(12)487 (13)338 (14)370 (15)269 (16)229
(17)329 (18)498 (19)507 (20)322 (21)286
(22)466 (23)399 (24)394 (25)425 (26)459
(27)381 (28)499 (29)410 (30)402

Page 29, Item 1:
(1)37 (2)4 (3)4 (4)14 (5)16 (6)4 (7)29 (8)27
(9)3 (10)27 (11)21 (12)46 (13)44 (14)16
(15)17 (16)23 (17)26 (18)5 (19)34 (20)27
(21)5 (22)29 (23)31 (24)38 (25)44 (26)12
(27)25 (28)40 (29)22 (30)41

Page 30, Item 1:
(1)9 (2)10 (3)11 (4)11 (5)13 (6)2 (7)15
(8)13 (9)14 (10)7 (11)6 (12)2 (13)18 (14)9
(15)40 (16)23 (17)0 (18)3 (19)0 (20)11
(21)16 (22)3 (23)22 (24)1 (25)14 (26)21
(27)36 (28)4 (29)34 (30)6

Page 31, Item 1:
(1)2 (2)9 (3)15 (4)3 (5)6 (6)4 (7)8 (8)1
(9)27 (10)13 (11)25 (12)3 (13)23 (14)38
(15)16 (16)34 (17)8 (18)25 (19)1 (20)0
(21)16 (22)8 (23)41 (24)11 (25)4 (26)15
(27)14 (28)39 (29)2 (30)26

Page 32, Item 1:
(1)20 (2)10 (3)18 (4)0 (5)30 (6)9 (7)15
(8)19 (9)1 (10)7 (11)27 (12)11 (13)25
(14)29 (15)6 (16)20 (17)25 (18)5 (19)34
(20)12 (21)25 (22)18 (23)18 (24)19 (25)4
(26)19 (27)18 (28)0 (29)7 (30)19

Page 33, Item 1:
(1)30 (2)18 (3)23 (4)21 (5)3 (6)28 (7)15
(8)7 (9)32 (10)9 (11)34 (12)21 (13)32
(14)16 (15)7 (16)19 (17)18 (18)4 (19)4
(20)16 (21)35 (22)6 (23)10 (24)8 (25)5
(26)35 (27)23 (28)39 (29)10 (30)14

Page 34, Item 1:
(1)27 (2)59 (3)11 (4)59 (5)23 (6)4 (7)6 (8)2
(9)37 (10)9 (11)60 (12)7 (13)35 (14)5
(15)31 (16)16 (17)32 (18)48 (19)33 (20)21
(21)19 (22)36 (23)5 (24)15 (25)34 (26)8
(27)4 (28)2 (29)6 (30)26

Page 35, Item 1:
(1)9 (2)80 (3)41 (4)14 (5)31 (6)23 (7)26
(8)26 (9)35 (10)101 (11)0 (12)0 (13)40
(14)55 (15)66 (16)31 (17)27 (18)75 (19)50
(20)38 (21)21 (22)2 (23)15 (24)52 (25)72
(26)41 (27)53 (28)44 (29)25 (30)20

Page 36, Item 1:
(1)6 (2)11 (3)75 (4)43 (5)24 (6)28 (7)76
(8)38 (9)1 (10)30 (11)2 (12)46 (13)14
(14)39 (15)1 (16)19 (17)44 (18)4 (19)55
(20)29 (21)75 (22)5 (23)17 (24)52 (25)21
(26)103 (27)36 (28)47 (29)69 (30)39

Page 37, Item 1:
(1)48 (2)121 (3)93 (4)34 (5)81 (6)44 (7)90
(8)78 (9)131 (10)24 (11)17 (12)17 (13)21
(14)94 (15)1 (16)10 (17)44 (18)73 (19)18
(20)58 (21)78 (22)127 (23)18 (24)63
(25)82 (26)21 (27)101 (28)32 (29)47
(30)60

Page 38, Item 1:
(1)30 (2)106 (3)81 (4)148 (5)119 (6)26
(7)91 (8)67 (9)110 (10)123 (11)77 (12)49
(13)45 (14)38 (15)25 (16)15 (17)104
(18)58 (19)23 (20)12 (21)72 (22)24
(23)144 (24)89 (25)40 (26)65 (27)36 (28)5
(29)19 (30)94

Page 39, Item 1:
(1)53 (2)38 (3)165 (4)135 (5)66 (6)224
(7)179 (8)105 (9)113 (10)17 (11)27 (12)58
(13)22 (14)102 (15)12 (16)25 (17)128
(18)77 (19)32 (20)62 (21)199 (22)94

(23)112 (24)60 (25)189 (26)40 (27)110
(28)121 (29)78 (30)109

Page 40, Item 1:
(1)36 (2)148 (3)64 (4)72 (5)18 (6)96 (7)65
(8)122 (9)49 (10)86 (11)74 (12)99 (13)149
(14)34 (15)43 (16)72 (17)12 (18)80 (19)1
(20)56 (21)20 (22)103 (23)87 (24)19
(25)99 (26)7 (27)18 (28)42 (29)89 (30)139

Page 41, Item 1:
(1)138 (2)87 (3)95 (4)73 (5)12 (6)186 (7)69
(8)17 (9)114 (10)93 (11)29 (12)13 (13)16
(14)178 (15)37 (16)49 (17)75 (18)15
(19)151 (20)132 (21)66 (22)106 (23)124
(24)149 (25)24 (26)68 (27)100 (28)54
(29)105 (30)19

Page 42, Item 1:
(1)28 (2)77 (3)13 (4)75 (5)170 (6)120 (7)45
(8)136 (9)1 (10)22 (11)163 (12)35 (13)15
(14)155 (15)54 (16)11 (17)0 (18)61 (19)86
(20)178 (21)30 (22)147 (23)20 (24)52
(25)38 (26)28 (27)119 (28)37 (29)121
(30)34

Page 43, Item 1:
(1)168 (2)188 (3)171 (4)28 (5)24 (6)10
(7)199 (8)107 (9)96 (10)111 (11)182
(12)32 (13)149 (14)87 (15)61 (16)102
(17)0 (18)43 (19)239 (20)51 (21)80 (22)92
(23)51 (24)81 (25)59 (26)99 (27)29 (28)86
(29)183 (30)52

Page 44, Item 1:
(1)104 (2)20 (3)75 (4)101 (5)97 (6)1 (7)26
(8)11 (9)19 (10)79 (11)43 (12)12 (13)20
(14)115 (15)47 (16)83 (17)75 (18)43
(19)131 (20)20 (21)75 (22)21 (23)85
(24)114 (25)76 (26)9 (27)33 (28)33
(29)105 (30)85

Page 45, Item 1:
(1)1 (2)136 (3)14 (4)68 (5)12 (6)36 (7)196
(8)23 (9)174 (10)52 (11)9 (12)114 (13)95
(14)221 (15)107 (16)130 (17)86 (18)225
(19)3 (20)113 (21)115 (22)126 (23)46
(24)235 (25)234 (26)244 (27)24 (28)3
(29)163 (30)12

Page 46, Item 1:
(1)125 (2)14 (3)140 (4)114 (5)192 (6)193
(7)137 (8)38 (9)38 (10)32 (11)195 (12)49
(13)58 (14)145 (15)264 (16)305 (17)61
(18)226 (19)157 (20)139 (21)110 (22)281
(23)12 (24)52 (25)37 (26)13 (27)85
(28)122 (29)79 (30)72

Page 47, Item 1:
(1)19 (2)13 (3)206 (4)152 (5)179 (6)336
(7)52 (8)142 (9)99 (10)259 (11)188 (12)15
(13)147 (14)306 (15)51 (16)258 (17)61
(18)24 (19)210 (20)106 (21)210 (22)49
(23)343 (24)260 (25)49 (26)244 (27)214
(28)45 (29)129 (30)123

Page 48, Item 1:
(1)113 (2)128 (3)371 (4)157 (5)90 (6)217
(7)7 (8)4 (9)254 (10)81 (11)32 (12)425
(13)132 (14)268 (15)333 (16)338 (17)392
(18)348 (19)205 (20)143 (21)181 (22)83
(23)348 (24)232 (25)18 (26)209 (27)121
(28)420 (29)202 (30)209

Page 50, Item 1:
(1)48 (2)8 (3)12 (4)63 (5)15 (6)24 (7)28
(8)14 (9)18 (10)42 (11)10 (12)21 (13)12
(14)12 (15)56 (16)48 (17)12 (18)42 (19)49
(20)54 (21)12 (22)24 (23)24 (24)72 (25)45
(26)42 (27)36 (28)63 (29)49 (30)56

Page 51, Item 1:
(1)21 (2)56 (3)28 (4)12 (5)48 (6)36 (7)81
(8)18 (9)42 (10)18 (11)42 (12)6 (13)15
(14)8 (15)48 (16)12 (17)27 (18)49 (19)32
(20)24 (21)21 (22)12 (23)21 (24)32 (25)64
(26)56 (27)12 (28)28 (29)28 (30)64

Page 52, Item 1:
(1)18 (2)20 (3)30 (4)54 (5)72 (6)10 (7)72
(8)28 (9)12 (10)20 (11)63 (12)36 (13)12
(14)30 (15)49 (16)10 (17)40 (18)56 (19)48
(20)81 (21)32 (22)45 (23)10 (24)30 (25)56
(26)16 (27)40 (28)63 (29)72 (30)56

Page 53, Item 1:
(1)72 (2)105 (3)33 (4)10 (5)70 (6)20 (7)90
(8)40 (9)65 (10)6 (11)30 (12)12 (13)4
(14)32 (15)75 (16)72 (17)26 (18)98 (19)8
(20)39 (21)40 (22)60 (23)32 (24)26 (25)56
(26)60 (27)36 (28)9 (29)14 (30)32

Page 54, Item 1:
(1)160 (2)81 (3)80 (4)105 (5)63 (6)76 (7)36
(8)45 (9)15 (10)38 (11)91 (12)135 (13)15
(14)21 (15)32 (16)76 (17)65 (18)140
(19)75 (20)36 (21)22 (22)36 (23)126
(24)108 (25)171 (26)30 (27)21 (28)36
(29)119 (30)54

Page 55, Item 1:
(1)140 (2)75 (3)78 (4)56 (5)189 (6)96
(7)145 (8)87 (9)91 (10)189 (11)27 (12)78
(13)133 (14)234 (15)40 (16)95 (17)60
(18)63 (19)36 (20)182 (21)232 (22)216
(23)87 (24)40 (25)54 (26)66 (27)144

(28)87 (29)252 (30)70

Page 56, Item 1:
(1)294 (2)490 (3)210 (4)98 (5)234 (6)30
(7)78 (8)90 (9)162 (10)144 (11)288
(12)148 (13)90 (14)294 (15)294 (16)78
(17)196 (18)350 (19)164 (20)305 (21)276
(22)186 (23)185 (24)288 (25)57 (26)216
(27)156 (28)125 (29)248 (30)259

Page 57, Item 1:
(1)344 (2)208 (3)27 (4)666 (5)198 (6)288
(7)195 (8)288 (9)88 (10)264 (11)72
(12)150 (13)333 (14)432 (15)144 (16)438
(17)468 (18)189 (19)315 (20)54 (21)212
(22)544 (23)171 (24)560 (25)385 (26)312
(27)148 (28)432 (29)348 (30)224

Page 58, Item 1:
(1)747 (2)120 (3)116 (4)712 (5)120 (6)258
(7)216 (8)147 (9)30 (10)210 (11)165
(12)244 (13)164 (14)630 (15)581 (16)177
(17)76 (18)195 (19)297 (20)180 (21)51
(22)696 (23)216 (24)175 (25)560 (26)378
(27)308 (28)272 (29)82 (30)154

Page 59, Item 1:
(1)720 (2)574 (3)343 (4)380 (5)235 (6)217
(7)450 (8)414 (9)480 (10)792 (11)368
(12)410 (13)513 (14)168 (15)238 (16)480
(17)280 (18)425 (19)392 (20)686 (21)448
(22)335 (23)440 (24)801 (25)609 (26)756
(27)432 (28)425 (29)224 (30)427

Page 60, Item 1:
(1)1350 (2)837 (3)200 (4)420 (5)973
(6)208 (7)231 (8)192 (9)324 (10)1001
(11)912 (12)968 (13)369 (14)520 (15)882
(16)340 (17)105 (18)416 (19)528 (20)693
(21)290 (22)720 (23)1208 (24)414 (25)655
(26)99 (27)750 (28)322 (29)512 (30)248
(31)770 (32)1120 (33)798 (34)178 (35)376
(36)1152

Page 61, Item 1:
(1)1616 (2)411 (3)1488 (4)320 (5)935
(6)1520 (7)1218 (8)1647 (9)1320 (10)882
(11)364 (12)345 (13)160 (14)1596 (15)992
(16)459 (17)1992 (18)1200 (19)772
(20)450 (21)306 (22)459 (23)948 (24)846
(25)1485 (26)438 (27)1398 (28)496
(29)296 (30)736 (31)1547 (32)508
(33)1008 (34)378 (35)708 (36)567

Page 62, Item 1:
(1)1550 (2)960 (3)960 (4)468 (5)1312
(6)1384 (7)232 (8)2000 (9)1631 (10)507
(11)1251 (12)704 (13)1404 (14)674
(15)1752 (16)975 (17)1036 (18)2176
(19)2280 (20)768 (21)756 (22)2712
(23)636 (24)870 (25)2768 (26)1666
(27)627 (28)990 (29)728 (30)1100
(31)1784 (32)1030 (33)1666 (34)1185
(35)1106 (36)1436

Page 63, Item 1:
(1)4320 (2)4172 (3)1724 (4)1278 (5)2115
(6)1940 (7)2499 (8)2580 (9)2175 (10)1386
(11)4815 (12)796 (13)1524 (14)1865
(15)1911 (16)2675 (17)2020 (18)1820
(19)1890 (20)1720 (21)3498 (22)3688
(23)3339 (24)1582 (25)714 (26)1371
(27)2292 (28)1136 (29)806 (30)4968
(31)1465 (32)1362 (33)2000 (34)4500

(35)1625 (36)1362

Page 64, Item 1:
(1)126 (2)290 (3)240 (4)340 (5)253 (6)228
(7)264 (8)350 (9)384 (10)360 (11)275
(12)225 (13)256 (14)80 (15)324 (16)198
(17)250 (18)198 (19)228 (20)352 (21)117
(22)110 (23)170 (24)288 (25)288 (26)121
(27)360 (28)324 (29)407 (30)408 (31)385
(32)220 (33)304 (34)539 (35)297 (36)88

Page 65, Item 1:
(1)216 (2)297 (3)410 (4)720 (5)804 (6)234
(7)594 (8)423 (9)567 (10)320 (11)333
(12)660 (13)636 (14)384 (15)408 (16)319
(17)460 (18)432 (19)828 (20)390 (21)234
(22)486 (23)638 (24)605 (25)310 (26)621
(27)242 (28)336 (29)539 (30)636

Page 66, Item 1:
(1)480 (2)1140 (3)935 (4)300 (5)341
(6)440 (7)564 (8)759 (9)220 (10)1056
(11)1116 (12)340 (13)408 (14)627
(15)1164 (16)473 (17)803 (18)828 (19)803
(20)360 (21)462 (22)230 (23)360 (24)950
(25)720 (26)781 (27)408 (28)444 (29)1067
(30)704

Page 67, Item 1:
(1)1872 (2)1474 (3)585 (4)1265 (5)1661
(6)2101 (7)960 (8)1890 (9)1386 (10)1413
(11)1464 (12)828 (13)1216 (14)1701
(15)1580 (16)756 (17)1408 (18)1788
(19)1600 (20)888 (21)640 (22)448

(23)1190 (24)1488 (25)1080 (26)2196
(27)968 (28)1314 (29)1958 (30)1560

Page 68, Item 1:
(1)1750 (2)2016 (3)4279 (4)3520 (5)3740
(6)1390 (7)1344 (8)2321 (9)4257 (10)1859
(11)4044 (12)2988 (13)1494 (14)1884
(15)2223 (16)2530 (17)2682 (18)1431
(19)2292 (20)2256 (21)1845 (22)3303
(23)927 (24)2691 (25)3186 (26)3204
(27)1150 (28)2655 (29)3396 (30)1782

Page 69, Item 1:
(1)4800 (2)6580 (3)4990 (4)4961 (5)7320
(6)8460 (7)5730 (8)5988 (9)6870 (10)6384
(11)6690 (12)8136 (13)7887 (14)9420
(15)5126 (16)7788 (17)7420 (18)7601
(19)6360 (20)8041 (21)7788 (22)8604
(23)5258 (24)5904 (25)6876 (26)6050
(27)5964 (28)5568 (29)6024 (30)6280

Page 71, Item 1:

(1)
```
      5
5 ) 2 5
  - 2 5
      0
```
(2)
```
      7
8 ) 5 6
  - 5 6
      0
```
(3)
```
      3
2 ) 6
  -   6
      0
```

(4)
```
      6
8 ) 4 8
  - 4 8
      0
```
(5)
```
      4
8 ) 3 2
  - 3 2
      0
```
(6)
```
      1 9
2 ) 3 8
  - 2
    1 8
  - 1 8
      0
```

(7)
```
      7
6 ) 4 2
  - 4 2
      0
```
(8)
```
      7
2 ) 1 4
  - 1 4
      0
```
(9)
```
      1
2 ) 2
  -   2
      0
```

(10)
```
      4
4 ) 1 6
  - 1 6
      0
```
(11)
```
      5
2 ) 1 0
  - 1 0
      0
```
(12)
```
      2
7 ) 1 4
  - 1 4
      0
```

(13)
```
      1 6
3 ) 4 8
  - 3
    1 8
  - 1 8
      0
```
(14)
```
      5
5 ) 2 5
  - 2 5
      0
```
(15)
```
      1
8 ) 8
  -   8
      0
```

(16)
```
      9
6 ) 5 4
  - 5 4
      0
```
(17)
```
      3
5 ) 1 5
  - 1 5
      0
```
(18)
```
      6
6 ) 3 6
  - 3 6
      0
```

(19)
```
      1
6 ) 6
  -   6
      0
```
(20)
```
      1 1
5 ) 5 5
  - 5
    0 5
  - 5
    0
```
(21)
```
      8
5 ) 4 0
  - 4 0
      0
```

(22)
```
      5
3 ) 1 5
  - 1 5
      0
```
(23)
```
      1 1
3 ) 3 3
  - 3
    0 3
  - 3
    0
```
(24)
```
      1 1
2 ) 2 2
  - 2
    0 2
  - 2
    0
```

(25)
```
      5
9 ) 4 5
  - 4 5
      0
```
(26)
```
      4
3 ) 1 2
  - 1 2
      0
```
(27)
```
      1 4
2 ) 2 8
  - 2
    0 8
  - 8
    0
```

(28)
```
      1 0
2 ) 2 0
  - 2
    0 0
  - 0
    0
```
(29)
```
      1
2 ) 2
  -   2
      0
```
(30)
```
      2
6 ) 1 2
  - 1 2
      0
```

Page 72, Item 1:

(1)
```
      1 7
3 ) 5 1
  - 3
    2 1
  - 2 1
      0
```
(2)
```
      2 0
4 ) 8 0
  - 8
    0 0
  -   0
      0
```
(3)
```
      1 0
4 ) 4 0
  - 4
    0 0
  -   0
      0
```

(4)
```
      3
4 ) 1 2
  - 1 2
      0
```
(5)
```
      1 4
4 ) 5 6
  - 4
    1 6
  - 1 6
      0
```
(6)
```
      8
8 ) 6 4
  - 6 4
      0
```

(7)
```
      1 9
3 ) 5 7
  - 3
    2 7
  - 2 7
      0
```
(8)
```
      1
8 ) 8
  -   8
      0
```
(9)
```
      1 9
3 ) 5 7
  - 3
    2 7
  - 2 7
      0
```

(10)
```
      3
3 ) 9
  -   9
      0
```
(11)
```
      1 7
4 ) 6 8
  - 4
    2 8
  - 2 8
      0
```
(12)
```
      3
0 ) 9
  -   9
      0
```

120

(15)
```
      2
7 ) 1 4
  - 1 4
      0
```
(16)
```
      3
3 ) 9
  -   9
      0
```
(17)
```
        3
8 ) 2 4
  - 2 4
      0
```
(4)
```
      1 6
5 ) 8 0
  - 5
    3 0
  - 3 0
      0
```
(5)
```
      1 3
5 ) 6 5
  - 5
    1 5
  - 1 5
      0
```
(6)
```
        8
4 ) 3 2
  - 3 2
      0
```

(18)
```
      1 4
3 ) 4 2
  - 3
    1 2
  - 1 2
      0
```
(19)
```
        5
7 ) 3 5
  - 3 5
      0
```
(20)
```
      1 5
5 ) 7 5
  - 5
    2 5
  - 2 5
      0
```
(7)
```
      2 9
3 ) 8 7
  - 6
    2 7
  - 2 7
      0
```
(8)
```
      1 9
3 ) 5 7
  - 3
    2 7
  - 2 7
      0
```
(9)
```
        7
8 ) 5 6
  - 5 6
      0
```

(21)
```
      1 9
4 ) 7 6
  - 4
    3 6
  - 3 6
      0
```
(22)
```
      1 0
8 ) 8 0
  - 8
    0 0
  -   0
      0
```
(23)
```
      1 5
3 ) 4 5
  - 3
    1 5
  - 1 5
      0
```
(10)
```
      1 5
3 ) 4 5
  - 3
    1 5
  - 1 5
      0
```
(11)
```
      1 0
3 ) 3 0
  - 3
    0 0
  -   0
      0
```
(12)
```
      1 1
9 ) 9 9
  - 9
    0 9
  -   9
      0
```

(24)
```
      1 3
6 ) 7 8
  - 6
    1 8
  - 1 8
      0
```
(25)
```
      2 6
3 ) 7 8
  - 6
    1 8
  - 1 8
      0
```
(26)
```
        2
4 ) 8
  -   8
      0
```
(13)
```
        9
8 ) 7 2
  - 7 2
      0
```
(14)
```
        7
5 ) 3 5
  - 3 5
      0
```
(15)
```
      1 4
5 ) 7 0
  - 5
    2 0
  - 2 0
      0
```

(27)
```
        4
4 ) 1 6
  - 1 6
      0
```
(28)
```
      1 1
7 ) 7 7
  - 7
    0 7
  -   7
      0
```
(29)
```
      1 1
5 ) 5 5
  - 5
    0 5
  -   5
      0
```
(16)
```
      1 0
3 ) 3 0
  - 3
    0 0
  -   0
      0
```
(17)
```
      1 1
4 ) 4 4
  - 4
    0 4
  -   4
      0
```
(18)
```
        5
3 ) 1 5
  - 1 5
      0
```

(30)
```
        4
3 ) 1 2
  - 1 2
      0
```
(19)
```
      1 1
7 ) 7 7
  - 7
    0 7
  -   7
      0
```
(20)
```
      2 3
3 ) 6 9
  - 6
    0 9
  -   9
      0
```

Page 73, Item 1:

(1)
```
      1 6
5 ) 8 0
  - 5
    3 0
  - 3 0
      0
```
(2)
```
      1 9
4 ) 7 6
  - 4
    3 6
  - 3 6
      0
```
(3)
```
      1 3
3 ) 3 9
  - 3
    0 9
  -   9
      0
```

121

```
(21)    1 0     (22)    1 1     (23)       8
     3)3 0           5)5 5           6)4 8
      - 3            - 5            - 4 8
       0 0            0 5              0
      -  0           -  5
         0              0
```

```
(24)    1 2     (25)      7     (26)    2 3
     8)9 6           7)4 9           3)6 9
      - 8            - 4 9          - 6
       1 6             0             0 9
      -1 6                          -  9
         0                             0
```

```
(27)    1 7     (28)    1 8     (29)    1 5
     3)5 1           4)7 2           4)6 0
      - 3            - 4            - 4
       2 1            3 2            2 0
      -2 1           -3 2           -2 0
         0              0              0
```

```
(30)      3
     4)1 2
      -1 2
         0
```

Page 74, Item 1:

```
(1)    1 4     (2)      9     (3)       3
    6)8 4          5)4 5          8)2 4
     - 6           - 4 5          - 2 4
      2 1            0              0
     -2 4
        0
```

```
(4)      7     (5)    3 3     (6)    1 8
    7)4 9          3)9 9          5)9 0
     - 4 9          - 9           - 5
        0            0 9            4 0
                    -  9          - 4 0
                       0             0
```

```
(7)    1 3     (8)    1 2     (9)    1 0
    4)5 2          5)6 0          9)9 0
     - 4           - 5           - 9
      1 2           1 0            0 0
     -1 2          -1 0           -  0
        0             0              0
```

```
(10)    1 7     (11)    2 3     (12)    1 9
     5)8 5           4)9 2           3)5 7
      - 5            - 8            - 3
       3 5            1 2            2 7
      -3 5           -1 2           -2 7
         0              0              0
```

```
(13)    1 4     (14)    2 3     (15)    3 3
     5)7 0           4)9 2           3)9 9
      - 5            - 8            - 9
       2 0            1 2            0 9
      -2 0           -1 2           -  9
         0              0              0
```

```
(16)      5     (17)    1 9     (18)    1 3
     5)2 5           3)5 7           5)6 5
      -2 5           - 3            - 5
         0            2 7            1 5
                     -2 7           -1 5
                        0              0
```

```
(19)      6     (20)      4     (21)    1 9
     6)3 6           8)3 2           3)5 7
      -3 6           -3 2           - 3
         0              0            2 7
                                    -2 7
                                       0
```

```
(22)    1 3     (23)    1 3     (24)    1 1
     7)9 1           7)9 1           7)7 7
      - 7            - 7            - 7
       2 1            2 1            0 7
      -2 1           -2 1           -  7
         0              0              0
```

```
(25)    2 5     (26)    1 1     (27)    1 7
     3)7 5           5)5 5           5)8 5
      - 6            - 5            - 5
       1 5            0 5            3 5
      -1 5           -  5           -3 5
         0              0              0
```

```
(28)      5     (29)    1 1     (30)    2 4
     8)4 0           4)4 4           4)9 6
      -4 0           - 4            - 8
         0            0 4            1 6
                     -  4           -1 6
                        0              0
```

Page 75, Item 1:

(1)
```
      4
  8 ) 3 2
    - 3 2
        0
```
(2)
```
      1 5
  5 ) 7 5
    - 5
      2 5
    - 2 5
        0
```
(3)
```
        4
  9 ) 3 6
    - 3 6
        0
```

(4)
```
        9
  5 ) 4 5
    - 4 5
        0
```
(5)
```
      1 4
  6 ) 8 4
    - 6
      2 4
    - 2 4
        0
```
(6)
```
        9
  8 ) 7 2
    - 7 2
        0
```

(7)
```
        3
  7 ) 2 1
    - 2 1
        0
```
(8)
```
        9
  9 ) 8 1
    - 8 1
        0
```
(9)
```
        2
  7 ) 1 4
    - 1 4
        0
```

(10)
```
        2
  8 ) 1 6
    - 1 6
        0
```
(11)
```
      1 4
  5 ) 7 0
    - 5
      2 0
    - 2 0
        0
```
(12)
```
      1 3
  5 ) 6 5
    - 5
      1 5
    - 1 5
        0
```

(13)
```
      1 3
  6 ) 7 8
    - 6
      1 8
    - 1 8
        0
```
(14)
```
        6
  9 ) 5 4
    - 5 4
        0
```
(15)
```
        6
  5 ) 3 0
    - 3 0
        0
```

(16)
```
        5
  7 ) 3 5
    - 3 5
        0
```
(17)
```
        9
  9 ) 8 1
    - 8 1
        0
```
(18)
```
        5
  5 ) 2 5
    - 2 5
        0
```

(19)
```
      1 7
  5 ) 8 5
    - 5
      3 5
    - 3 5
        0
```
(20)
```
        8
  5 ) 4 0
    - 4 0
        0
```
(21)
```
        7
  7 ) 4 9
    - 4 9
        0
```

(22)
```
        6
  8 ) 4 8
    - 4 8
        0
```
(23)
```
      1 8
  5 ) 9 0
    - 5
      4 0
    - 4 0
        0
```
(24)
```
        7
  5 ) 3 5
    - 3 5
        0
```

(25)
```
      1 1
  9 ) 9 9
    - 9
      0 9
    - 9
      0
```
(26)
```
        6
  8 ) 4 8
    - 4 8
        0
```
(27)
```
      1 3
  7 ) 9 1
    - 7
      2 1
    - 2 1
        0
```

(28)
```
        4
  7 ) 2 8
    - 2 8
        0
```
(29)
```
      1 4
  6 ) 8 4
    - 6
      2 4
    - 2 4
        0
```
(30)
```
      1 3
  6 ) 7 8
    - 6
      1 8
    - 1 8
        0
```

Page 76, Item 1:

(1)
```
      2 8
  3 ) 8 4
    - 6
      2 4
    - 2 4
        0
```
(2)
```
      2 9
  3 ) 8 7
    - 6
      2 7
    - 2 7
        0
```
(3)
```
      1 1
  5 ) 5 5
    - 5
      0 5
    - 5
        0
```

(4)
```
        7
  5 ) 3 5
    - 3 5
        0
```
(5)
```
      1 5
  3 ) 4 5
    - 3
      1 5
    - 1 5
        0
```
(6)
```
      2 5
  3 ) 7 5
    - 6
      1 5
    - 1 5
        0
```

(7)
```
      1 4
  5 ) 7 0
    - 5
      2 0
    - 2 0
        0
```
(8)
```
      1 5
  5 ) 7 5
    - 5
      2 5
    - 2 5
        0
```
(9)
```
      1 6
  6 ) 9 6
    - 6
      3 6
    - 3 6
        0
```

(10)
```
      1 1
  5 ) 5 5
    - 5
      0 5
    - 5
        0
```
(11)
```
        2
  5 ) 1 0
    - 1 0
        0
```

123

(12)
```
     1 2
  8 ) 9 6
   -  8
      1 6
   -  1 6
        0
```

(13)
```
       3
  6 ) 1 8
   - 1 8
       0
```

(14)
```
     1 4
  7 ) 9 8
   -  7
      2 8
   -  2 8
        0
```

(15)
```
       9
  9 ) 8 1
   - 8 1
       0
```

(16)
```
     1 7
  4 ) 6 8
   -  4
      2 8
   -  2 8
        0
```

(17)
```
       4
  4 ) 1 6
   - 1 6
       0
```

(18)
```
       2
  6 ) 1 2
   - 1 2
       0
```

(19)
```
     1 6
  6 ) 9 6
   -  6
      3 6
   -  3 6
        0
```

(20)
```
     1 4
  3 ) 4 2
   -  3
      1 2
   -  1 2
        0
```

(21)
```
     1 8
  3 ) 5 4
   -  3
      2 4
   -  2 4
        0
```

(22)
```
       2
  6 ) 1 2
   - 1 2
       0
```

(23)
```
     2 9
  3 ) 8 7
   -  6
      2 7
   -  2 7
        0
```

(24)
```
       4
  5 ) 2 0
   - 2 0
       0
```

(25)
```
     1 2
  4 ) 4 8
   -  4
      0 8
   -    8
        0
```

(26)
```
       2
  5 ) 1 0
   - 1 0
       0
```

(27)
```
     1 0
  8 ) 8 0
   -  8
      0 0
   -    0
        0
```

(28)
```
     1 1
  8 ) 8 8
   -  8
      0 8
   -    8
        0
```

(29)
```
       5
  6 ) 3 0
   - 3 0
       0
```

(30)
```
       4
  6 ) 2 4
   - 2 4
       0
```

Page 77, Item 1:

(1)
```
       2 4
  6 ) 1 4 4
   -  1 2
        2 4
   -    2 4
          0
```

(2)
```
       3 1
  3 ) 9 3
   -  9
        0 3
   -      3
          0
```

(3)
```
       1 1
  3 ) 3 3
   -  3
        0 3
   -      3
          0
```

(4)
```
       1 7
  3 ) 5 1
   -  3
        2 1
   -    2 1
          0
```

(5)
```
       1 5
  5 ) 7 5
   -  5
        2 5
   -    2 5
          0
```

(6)
```
         3 0
  5 ) 1 5 0
   -  1 5
          0 0
   -        0
            0
```

(7)
```
         5
  6 ) 3 0
   -  3 0
          0
```

(8)
```
       2 9
  4 ) 1 1 6
   -    8
        3 6
   -    3 6
          0
```

(9)
```
       1 2
  7 ) 8 4
   -  7
        1 4
   -    1 4
          0
```

(10)
```
         4
  4 ) 1 6
   -  1 6
          0
```

(11)
```
         1 5
  8 ) 1 2 0
   -    8
          4 0
   -      4 0
            0
```

(12)
```
       2 3
  3 ) 6 9
   -  6
        0 9
   -      9
          0
```

(13)
```
       2 6
  3 ) 7 8
   -  6
        1 8
   -    1 8
          0
```

(14)
```
         1 4
  5 ) 7 0
   -    5
          2 0
   -      2 0
            0
```

(15)
```
         1 3
  6 ) 7 8
   -    6
          1 8
   -      1 8
            0
```

(16)
```
         1 9
  3 ) 5 7
   -    3
          2 7
   -      2 7
            0
```

(17)
```
           3
  9 ) 2 7
   -    2 7
            0
```

(18)
```
         2 0
  6 ) 1 2 0
   -    1 2
            0 0
   -          0
```

(19)
```
      2 4
  4 ) 9 6
    - 8
      1 6
    - 1 6
        0
```
(20)
```
      1 7
  5 ) 8 5
    - 5
      3 5
    - 3 5
        0
```
(12)
```
      1 0
  5 ) 5 0
    - 5
      0 0
    - 0
        0
```
(13)
```
      2 9
  3 ) 8 7
    - 6
      2 7
    - 2 7
        0
```

(21)
```
      2 3
  4 ) 9 2
    - 8
      1 2
    - 1 2
        0
```
(22)
```
      1 3
  3 ) 3 9
    - 3
      0 9
    - 9
        0
```
(14)
```
      2 3
  6 ) 1 3 8
    - 1 2
        1 8
      - 1 8
          0
```
(15)
```
      1 3
  4 ) 5 2
    - 4
      1 2
    - 1 2
        0
```

(23)
```
      2 3
  4 ) 9 2
    - 8
      1 2
    - 1 2
        0
```
(24)
```
      2 6
  5 ) 1 3 0
    - 1 0
        3 0
      - 3 0
          0
```
(16)
```
      3 7
  4 ) 1 4 8
    - 1 2
        2 8
      - 2 8
          0
```
(17)
```
      1 1
  4 ) 4 4
    - 4
      0 4
    - 4
        0
```

Page 78, Item 1:

(1)
```
      1 2
  6 ) 7 2
    - 6
      1 2
    - 1 2
        0
```
(2)
```
        8
  8 ) 6 4
    - 6 4
        0
```
(3)
```
      3 9
  4 ) 1 5 6
    - 1 2
        3 6
      - 3 6
          0
```

(18)
```
        8
  8 ) 6 4
    - 6 4
        0
```
(19)
```
      1 4
  3 ) 4 2
    - 3
      1 2
    - 1 2
        0
```

(4)
```
      2 9
  4 ) 1 1 6
    - 8
      3 6
    - 3 6
        0
```
(5)
```
        7
  7 ) 4 9
    - 4 9
        0
```
(6)
```
      2 5
  5 ) 1 2 5
    - 1 0
        2 5
      - 2 5
          0
```

(20)
```
      1 1
  8 ) 8 8
    - 8
      0 8
    - 8
        0
```
(21)
```
      2 7
  6 ) 1 6 2
    - 1 2
        4 2
      - 4 2
          0
```

(7)
```
      2 2
  5 ) 1 1 0
    - 1 0
        1 0
      - 1 0
          0
```
(8)
```
        5
  7 ) 3 5
    - 3 5
        0
```
(9)
```
      1 5
  9 ) 1 3 5
    - 9
      4 5
    - 4 5
        0
```

(22)
```
      3 2
  3 ) 9 6
    - 9
      0 6
    - 6
        0
```
(23)
```
      1 3
  4 ) 5 2
    - 4
      1 2
    - 1 2
        0
```

(10)
```
      1 3
  4 ) 5 2
    - 4
      1 2
    - 1 2
        0
```
(11)
```
      6 2
  3 ) 1 8 6
    - 1 8
        0 6
      - 6
          0
```

(24)
```
      3 8
  4 ) 1 5 2
    - 1 2
        3 2
      - 3 2
          0
```

(7)
$$\begin{array}{r} 31 \\ 7\overline{)217} \\ -21 \\ \hline 07 \\ -7 \\ \hline 0 \end{array}$$

(8)
$$\begin{array}{r} 133 \\ 3\overline{)399} \\ -3 \\ \hline 09 \\ -9 \\ \hline 09 \\ -9 \\ \hline 0 \end{array}$$

(9)
$$\begin{array}{r} 34 \\ 7\overline{)238} \\ -21 \\ \hline 28 \\ -28 \\ \hline 0 \end{array}$$

(20)
$$\begin{array}{r} 31 \\ 7\overline{)217} \\ -21 \\ \hline 07 \\ -7 \\ \hline 0 \end{array}$$

(21)
$$\begin{array}{r} 50 \\ 7\overline{)350} \\ -35 \\ \hline 00 \\ -0 \\ \hline 0 \end{array}$$

(10)
$$\begin{array}{r} 16 \\ 8\overline{)128} \\ -8 \\ \hline 48 \\ -48 \\ \hline 0 \end{array}$$

(11)
$$\begin{array}{r} 38 \\ 7\overline{)266} \\ -21 \\ \hline 56 \\ -56 \\ \hline 0 \end{array}$$

(22)
$$\begin{array}{r} 41 \\ 4\overline{)164} \\ -16 \\ \hline 04 \\ -4 \\ \hline 0 \end{array}$$

(23)
$$\begin{array}{r} 28 \\ 5\overline{)140} \\ -10 \\ \hline 40 \\ -40 \\ \hline 0 \end{array}$$

(12)
$$\begin{array}{r} 33 \\ 5\overline{)165} \\ -15 \\ \hline 15 \\ -15 \\ \hline 0 \end{array}$$

(13)
$$\begin{array}{r} 53 \\ 7\overline{)371} \\ -35 \\ \hline 21 \\ -21 \\ \hline 0 \end{array}$$

(24)
$$\begin{array}{r} 29 \\ 3\overline{)87} \\ -6 \\ \hline 27 \\ -27 \\ \hline 0 \end{array}$$

(14)
$$\begin{array}{r} 46 \\ 7\overline{)322} \\ -28 \\ \hline 42 \\ -42 \\ \hline 0 \end{array}$$

(15)
$$\begin{array}{r} 30 \\ 3\overline{)90} \\ -9 \\ \hline 00 \\ -0 \\ \hline 0 \end{array}$$

(16)
$$\begin{array}{r} 36 \\ 9\overline{)324} \\ -27 \\ \hline 54 \\ -54 \\ \hline 0 \end{array}$$

(17)
$$\begin{array}{r} 104 \\ 3\overline{)312} \\ -3 \\ \hline 01 \\ -0 \\ \hline 12 \\ -12 \\ \hline 0 \end{array}$$

(18)
$$\begin{array}{r} 14 \\ 8\overline{)112} \\ -8 \\ \hline 32 \\ -32 \\ \hline 0 \end{array}$$

(19)
$$\begin{array}{r} 77 \\ 3\overline{)231} \\ -21 \\ \hline 21 \\ -21 \\ \hline 0 \end{array}$$

Page 80, Item 1:

(1)
$$\begin{array}{r} 183 \\ 2\overline{)366} \\ -2 \\ \hline 16 \\ -16 \\ \hline 06 \\ -6 \\ \hline 0 \end{array}$$

(2)
$$\begin{array}{r} 68 \\ 4\overline{)272} \\ -24 \\ \hline 32 \\ -32 \\ \hline 0 \end{array}$$

(3)
$$\begin{array}{r} 35 \\ 5\overline{)175} \\ -15 \\ \hline 25 \\ -25 \\ \hline 0 \end{array}$$

(4)
$$\begin{array}{r} 75 \\ 5\overline{)375} \\ -35 \\ \hline 25 \\ -25 \\ \hline 0 \end{array}$$

(5)
$$\begin{array}{r} 79 \\ 5\overline{)395} \\ -35 \\ \hline 45 \\ -45 \\ \hline 0 \end{array}$$

(6)
$$\begin{array}{r} 19 \\ 7\overline{)133} \\ -7 \\ \hline 63 \\ -63 \\ \hline 0 \end{array}$$

(10)
```
      1 2 4
  2 )2 4 8
   - 2
     0 4
   -   4
       0 8
   -     8
         0
```

(11)
```
        2 0
  9 )1 8 0
   - 1 8
       0 0
   -     0
         0
```

(22)
```
        2 7
  6 )1 6 2
   - 1 2
       4 2
   -   4 2
         0
```

(23)
```
        6 3
  6 )3 7 8
   - 3 6
       1 8
   -   1 8
         0
```

(12)
```
      1 6 2
  3 )4 8 6
   - 3
     1 8
   - 1 8
       0 6
   -     6
         0
```

(13)
```
      1 4 7
  2 )2 9 4
   - 2
     0 9
   -   8
       1 4
   -   1 4
         0
```

(24)
```
        2 7
  7 )1 8 9
   - 1 4
       4 9
   -   4 9
         0
```

(14)
```
      1 5 4
  2 )3 0 8
   - 2
     1 0
   - 1 0
       0 8
   -     8
         0
```

(15)
```
        3 9
  9 )3 5 1
   - 2 7
       8 1
   -   8 1
         0
```

Page 81, Item 1:

(1)
```
      1 5 5
  3 )4 6 5
   - 3
     1 6
   - 1 5
       1 5
   -   1 5
         0
```

(2)
```
        9 5
  3 )2 8 5
   - 2 7
       1 5
   -   1 5
         0
```

(3)
```
        5 5
  5 )2 7 5
   - 2 5
       2 5
   -   2 5
         0
```

(16)
```
        6 5
  2 )1 3 0
   - 1 2
       1 0
   -   1 0
         0
```

(17)
```
      1 1 3
  2 )2 2 6
   - 2
     0 2
   -   2
       0 6
   -     6
         0
```

(4)
```
        4 6
  7 )3 2 2
   - 2 8
       4 2
   -   4 2
         0
```

(5)
```
        6 8
  7 )4 7 6
   - 4 2
       5 6
   -   5 6
         0
```

(6)
```
        1 5
  3 )4 5
   - 3
     1 5
   - 1 5
       0
```

(18)
```
        9 3
  3 )2 7 9
   - 2 7
       0 9
   -     9
         0
```

(19)
```
        8 8
  3 )2 6 4
   - 2 4
       2 4
   -   2 4
         0
```

(7)
```
      1 4 3
  3 )4 2 9
   - 3
     1 2
   - 1 2
       0 9
   -     9
         0
```

(8)
```
        2 3
  7 )1 6 1
   - 1 4
       2 1
   -   2 1
         0
```

(9)
```
        2 5
  5 )1 2 5
   - 1 0
       2 5
   -   2 5
         0
```

(20)
```
        5 6
  8 )4 4 8
   - 4 0
       4 8
   -   4 8
         0
```

(21)
```
        6 6
  7 )4 6 2
   - 4 2
       4 2
   -   4 2
         0
```

(13)
```
    1 2 0
3 ) 3 6 0
  - 3
    0 6
  -   6
      0 0
  -     0
        0
```

(14)
```
      6 7
3 ) 2 0 1
  - 1 8
      2 1
    - 2 1
        0
```

(15)
```
    1 5 6
3 ) 4 6 8
  - 3
    1 6
  - 1 5
      1 8
    - 1 8
        0
```

(16)
```
      7 9
6 ) 4 7 4
  - 4 2
      5 4
    - 5 4
        0
```

(17)
```
    1 4 2
3 ) 4 2 6
  - 3
    1 2
  - 1 2
      0 6
    -   6
        0
```

(18)
```
    1 2 1
3 ) 3 6 3
  - 3
    0 6
  -   6
      0 3
    -   3
        0
```

(19)
```
      9 7
4 ) 3 8 8
  - 3 6
      2 8
    - 2 8
        0
```

(20)
```
      2 3
5 ) 1 1 5
  - 1 0
      1 5
    - 1 5
        0
```

(21)
```
    1 0 7
4 ) 4 2 8
  - 4
    0 2
  -   0
      2 8
    - 2 8
        0
```

(22)
```
      6 7
3 ) 2 0 1
  - 1 8
      2 1
    - 2 1
        0
```

(23)
```
      6 7
7 ) 4 6 9
  - 4 2
      4 9
    - 4 9
        0
```

(24)
```
      3 3
3 ) 9 9
  - 9
    0 9
  -   9
      0
```

Page 82, Item 1:

(1)
```
      1 4
8 ) 1 1 2
  -   8
      3 2
    - 3 2
        0
```

(2)
```
      2 6
7 ) 1 8 2
  - 1 4
      4 2
    - 4 2
        0
```

(3)
```
      8 4
4 ) 3 3 6
  - 3 2
      1 6
    - 1 6
        0
```

(4)
```
        5 4
8 ) 4 3 2
  - 4 0
      3 2
    - 3 2
        0
```

(5)
```
        4
4 ) 1 6
  - 1 6
      0
```

(6)
```
      1 0 9
4 ) 4 3 6
  - 4
    0 3
  -   0
      3 6
    - 3 6
        0
```

(7)
```
        1 1
8 ) 8 8
  - 8
    0 8
  -   8
      0
```

(8)
```
      1 5 7
3 ) 4 7 1
  - 3
    1 7
  - 1 5
      2 1
    - 2 1
        0
```

(9)
```
        9 1
4 ) 3 6 4
  - 3 6
      0 4
    -   4
        0
```

(10)
```
        2 6
8 ) 2 0 8
  - 1 6
      4 8
    - 4 8
        0
```

(11)
```
        9 7
5 ) 4 8 5
  - 4 5
      3 5
    - 3 5
        0
```

(12)
```
      1 3 2
3 ) 3 9 6
  - 3
    0 9
  -   9
      0 6
    -   6
        0
```

(13)
```
        3 7
5 ) 1 8 5
  - 1 5
      3 5
    - 3 5
        0
```

128

(16)
```
        3 5
  7 ) 2 4 5
    - 2 1
        3 5
      - 3 5
          0
```
(17)
```
        8 0
  3 ) 2 4 0
    - 2 4
        0 0
      -   0
          0
```

(18)
```
        2 6
  5 ) 1 3 0
    - 1 0
        3 0
      - 3 0
          0
```
(19)
```
        7 4
  3 ) 2 2 2
    - 2 1
        1 2
      - 1 2
          0
```

(20)
```
        3 8
  7 ) 2 6 6
    - 2 1
        5 6
      - 5 6
          0
```
(21)
```
        2 7
  9 ) 2 4 3
    - 1 8
        6 3
      - 6 3
          0
```

(22)
```
      1 3 1
  3 ) 3 9 3
    - 3
      0 9
    -   9
        0 3
      -   3
          0
```
(23)
```
        5 8
  7 ) 4 0 6
    - 3 5
        5 6
      - 5 6
          0
```

(24)
```
        3 3
  5 ) 1 6 5
    - 1 5
        1 5
      - 1 5
          0
```

Page 83, Item 1:

(1)
```
      2 4
  3 ) 7 2
    - 6
      1 2
    - 1 2
        0
```
(2)
```
        9
  9 ) 8 1
    - 8 1
        0
```
(3)
```
      1 1
  5 ) 5 5
    - 5
      0 5
    -   5
        0
```

(4)
```
        3
  1 2 ) 3 6
      - 3 6
          0
```
(5)
```
        7
  9 ) 6 3
    - 6 3
        0
```
(6)
```
        4
  5 ) 2 0
    - 2 0
        0
```

(7)
```
        2
  9 ) 1 8
    - 1 8
        0
```
(8)
```
          2
  1 0 ) 2 0
      - 2 0
          0
```
(9)
```
        7
  7 ) 4 9
    - 4 9
        0
```

(10)
```
        2
  7 ) 1 4
    - 1 4
        0
```
(11)
```
        9
  7 ) 6 3
    - 6 3
        0
```
(12)
```
      1 7
  4 ) 6 8
    - 4
      2 8
    - 2 8
        0
```

(13)
```
        5
  3 ) 1 5
    - 1 5
        0
```
(14)
```
      1 3
  7 ) 9 1
    - 7
      2 1
    - 2 1
        0
```
(15)
```
        4
  4 ) 1 6
    - 1 6
        0
```

(16)
```
        2
  8 ) 1 6
    - 1 6
        0
```
(17)
```
        3
  4 ) 1 2
    - 1 2
        0
```
(18)
```
        9
  7 ) 6 3
    - 6 3
        0
```

(19)
```
          5
  1 0 ) 5 0
      - 5 0
          0
```
(20)
```
        3
  7 ) 2 1
    - 2 1
        0
```
(21)
```
      1 1
  3 ) 3 3
    - 3
      0 3
    -   3
        0
```

(22)
```
          3
  1 0 ) 3 0
      - 3 0
          0
```
(23)
```
      1 1
  9 ) 9 9
    - 9
      0 9
    -   9
        0
```
(24)
```
      2 9
  3 ) 8 7
    - 6
      2 7
    - 2 7
        0
```

129

(9)
```
      7
  5)3 5
  - 3 5
      0
```
(10)
```
    3 1
  3)9 3
  - 9
    0 3
    - 3
      0
```
(11)
```
      3
  4)1 2
  - 1 2
      0
```
(4)
```
    5 3
  9)4 7 7
  - 4 5
      2 7
    - 2 7
        0
```
(5)
```
    9 7
  7)6 7 9
  - 6 3
      4 9
    - 4 9
        0
```
(6)
```
    1 9
  9)1 7 1
  - 9
    8 1
  - 8 1
      0
```
(12)
```
      9
  3)2 7
  - 2 7
      0
```
(13)
```
      6
  6)3 6
  - 3 6
      0
```
(14)
```
    1 7
  5)8 5
  - 5
    3 5
  - 3 5
      0
```
(7)
```
    8 6
  8)6 8 8
  - 6 4
      4 8
    - 4 8
        0
```
(8)
```
    3 4
  8)2 7 2
  - 2 4
      3 2
    - 3 2
        0
```
(9)
```
    9 4
  7)6 5 8
  - 6 3
      2 8
    - 2 8
        0
```
(15)
```
    3 1
  3)9 3
  - 9
    0 3
  - 3
    0
```
(16)
```
    1 1
  3)3 3
  - 3
    0 3
  - 3
    0
```
(17)
```
    1 1
  6)6 6
  - 6
    0 6
  - 6
    0
```
(10)
```
    7 9
  7)5 5 3
  - 4 9
      6 3
    - 6 3
        0
```
(11)
```
    5 9
  8)4 7 2
  - 4 0
      7 2
    - 7 2
        0
```
(18)
```
    1 6
  3)4 8
  - 3
    1 8
  - 1 8
      0
```
(19)
```
      6
  9)5 4
  - 5 4
      0
```
(20)
```
        9
  1 1)9 9
    - 9 9
        0
```
(12)
```
      1 1 6
  5)5 8 0
  - 5
    0 8
  -   5
      3 0
    - 3 0
        0
```
(13)
```
    6 4
  8)5 1 2
  - 4 8
      3 2
    - 3 2
        0
```
(21)
```
    1 1
  8)8 8
  - 8
    0 8
  - 8
    0
```
(22)
```
    2 8
  3)8 4
  - 6
    2 4
  - 2 4
      0
```
(23)
```
    2 3
  3)6 9
  - 6
    0 9
  - 9
    0
```
(14)
```
    2 1
  7)1 4 7
  - 1 4
      0 7
    -   7
        0
```
(15)
```
    3 2
  8)2 5 6
  - 2 4
      1 6
    - 1 6
        0
```
(24)
```
    1 9
  3)5 7
  - 3
    2 7
  - 2 7
      0
```

Page 85, Item 1:

(1)
```
      8 0
  8)6 4 0
  - 6 4
      0 0
    -   0
        0
```
(2)
```
    2 7
  8)2 1 6
  - 1 6
      5 6
    - 5 6
        0
```
(3)
```
    7 3
  6)4 3 8
  - 4 2
      1 8
    - 1 8
        0
```

130

(16)
```
      3 7
9 ) 3 3 3
  - 2 7
      6 3
    - 6 3
        0
```

(17)
```
      1 2 3
5 ) 6 1 5
  - 5
      1 1
    - 1 0
        1 5
      - 1 5
          0
```

(18)
```
      1 2 1
5 ) 6 0 5
  - 5
      1 0
    - 1 0
        0 5
      -   5
          0
```

(19)
```
      6 9
9 ) 6 2 1
  - 5 4
      8 1
    - 8 1
        0
```

(20)
```
      5 7
9 ) 5 1 3
  - 4 5
      6 3
    - 6 3
        0
```

(21)
```
      9 3
7 ) 6 5 1
  - 6 3
      2 1
    - 2 1
        0
```

(22)
```
      1 0 0
5 ) 5 0 0
  - 5
      0 0
    -   0
        0 0
      -   0
          0
```

(23)
```
      7 1
9 ) 6 3 9
  - 6 3
      0 9
    -   9
        0
```

(24)
```
      1 9
7 ) 1 3 3
  - 7
      6 3
    - 6 3
        0
```

Page 86, Item 1:

(1)
```
      1 7 3
3 ) 5 1 9
  - 3
      2 1
    - 2 1
        0 9
      -   9
          0
```

(2)
```
      1 3
4 ) 5 2
  - 4
      1 2
    - 1 2
        0
```

(3)
```
      5 3
6 ) 3 1 8
  - 3 0
      1 8
    - 1 8
        0
```

(4)
```
      4 9
4 ) 1 9 6
  - 1 6
      3 6
    - 3 6
        0
```

(5)
```
      5 9
8 ) 4 7 2
  - 4 0
      7 2
    - 7 2
        0
```

(6)
```
      5 7
6 ) 3 4 2
  - 3 0
      4 2
    - 4 2
        0
```

(7)
```
      5 9
4 ) 2 3 6
  - 2 0
      3 6
    - 3 6
        0
```

(8)
```
      3 9
3 ) 1 1 7
  - 9
      2 7
    - 2 7
        0
```

(9)
```
      6 6
5 ) 3 3 0
  - 3 0
      3 0
    - 3 0
        0
```

(10)
```
      2 3
3 ) 6 9
  - 6
      0 9
    -   9
        0
```

(11)
```
      6 0
7 ) 4 2 0
  - 4 2
      0 0
    -   0
        0
```

(12)
```
      2 7
9 ) 2 4 3
  - 1 8
      6 3
    - 6 3
        0
```

(13)
```
      3 5
6 ) 2 1 0
  - 1 8
      3 0
    - 3 0
        0
```

(14)
```
      9 2
7 ) 6 4 4
  - 6 3
      1 4
    - 1 4
        0
```

(15)
```
      8 9
6 ) 5 3 4
  - 4 8
      5 4
    - 5 4
        0
```

```
(19)        6 7    (20)        6 8
      8 ) 5 3 6          4 ) 2 7 2
        - 4 8              - 2 4
          5 6                3 2
        - 5 6              - 3 2
            0                  0

(21)        1 7    (22)        5 8
      3 ) 5 1            3 ) 1 7 4
        -   3              - 1 5
            2 1              2 4
        -   2 1            - 2 4
              0              0

(23)      1 6 4    (24)        9 8
      4 ) 6 5 6          7 ) 6 8 6
        - 4                - 6 3
          2 5                5 6
        - 2 4              - 5 6
            1 6                0
          - 1 6
              0
```

```
(10)        9      (11)      1 7    (12)      1 4
      5 ) 4 5          5 ) 8 5          7 ) 9 8
        - 4 5            - 5              - 7
            0              3 5              2 8
                         - 3 5            - 2 8
                             0                0

(13)      1 9      (14)        5    (15)          1
      5 ) 9 5          5 ) 2 5        1 1 ) 1 1
        - 5              - 2 5            - 1 1
          4 5                0                0
        - 4 5
            0

(16)      1 4      (17)        8    (18)          7
      6 ) 8 4        1 0 ) 8 0          9 ) 6 3
        - 6              - 8 0            - 6 3
          2 4                0                0
        - 2 4
            0

(19)        9      (20)        5    (21)          5
      6 ) 5 4          8 ) 4 0          8 ) 4 0
        - 5 4            - 4 0            - 4 0
            0                0                0

(22)      1 9      (23)        2    (24)          2
      5 ) 9 5          6 ) 1 2          8 ) 1 6
        - 5              - 1 2            - 1 6
          4 5                0                0
        - 4 5
            0
```

Page 87, Item 1:

```
(1)         7      (2)       1 9    (3)       1 3
      7 ) 4 9          5 ) 9 5          5 ) 6 5
        - 4 9            - 5              - 5
            0              4 5              1 5
                         - 4 5            - 1 5
                             0                0

(4)       1 8      (5)         5    (6)       1 6
      5 ) 9 0          9 ) 4 5          6 ) 9 6
        - 5              - 4 5            - 6
          4 0                0              3 6
        - 4 0                            - 3 6
            0                                0

(7)       1 3      (8)       1 9    (9)           8
      6 ) 7 8          5 ) 9 5        1 1 ) 8 8
        - 6              - 5              - 8 8
          1 8              4 5                0
        - 1 8            - 4 5
            0                0
```

Page 88, Item 1:

```
(1)       5 8      (2)     2 1 3    (3)     1 6 7
      9 ) 5 2 2        3 ) 6 3 9          3 ) 5 0 1
        - 4 5            - 6                - 3
          7 2              0 3                2 0
        - 7 2            -   3              - 1 8
            0                0 9                2 1
                         -   9              - 2 1
                             0                  0

(4)     1 8 7      (5)       8 8    (6)     2 3 0
      4 ) 7 4 8        4 ) 3 5 2          3 ) 6 9 0
        - 4              - 3 2              - 6
          3 4              3 2                0 9
        - 3 2            - 3 2              -   9
            2 8              0                  0 0
          - 2 8                             -     0
              0                                  0
```

132

(7)
```
    2 1 3
3 ) 6 3 9
  - 6
    0 3
  -   3
      0 9
    -   9
        0
```

(8)
```
    6 7
5 ) 3 3 5
  - 3 0
      3 5
    - 3 5
        0
```

(9)
```
    1 4 3
5 ) 7 1 5
  - 5
    2 1
  - 2 0
      1 5
    - 1 5
        0
```

(20)
```
      4 5
9 ) 4 0 5
  - 3 6
      4 5
    - 4 5
        0
```

(21)
```
    1 3 1
3 ) 3 9 3
  - 3
    0 9
  -   9
      0 3
    -   3
        0
```

(10)
```
    1 1 9
5 ) 5 9 5
  - 5
    0 9
  -   5
      4 5
    - 4 5
        0
```

(11)
```
    1 7 3
3 ) 5 1 9
  - 3
    2 1
  - 2 1
      0 9
    -   9
        0
```

(22)
```
    1 3 6
3 ) 4 0 8
  - 3
    1 0
  -   9
      1 8
    - 1 8
        0
```

(23)
```
      5 9
8 ) 4 7 2
  - 4 0
      7 2
    - 7 2
        0
```

(12)
```
    1 2 2
3 ) 3 6 6
  - 3
    0 6
  -   6
      0 6
    -   6
        0
```

(13)
```
        4 7
9 ) 4 2 3
  - 3 6
      6 3
    - 6 3
        0
```

(24)
```
    1 1 5
5 ) 5 7 5
  - 5
    0 7
  -   5
      2 5
    - 2 5
        0
```

(14)
```
        9 1
5 ) 4 5 5
  - 4 5
      0 5
    -   5
        0
```

(15)
```
    1 6 7
3 ) 5 0 1
  - 3
    2 0
  - 1 8
      2 1
    - 2 1
        0
```

Page 89, Item 1:

(1)
```
    3 0 7
3 ) 9 2 1
  - 9
    0 2
  -   0
      2 1
    - 2 1
        0
```

(2)
```
    1 2 6
7 ) 8 8 2
  - 7
    1 8
  - 1 4
      4 2
    - 4 2
        0
```

(3)
```
    1 0 0
5 ) 5 0 0
  - 5
    0 0
  -   0
      0 0
    -   0
        0
```

(16)
```
        7 7
6 ) 4 6 2
  - 4 2
      4 2
    - 4 2
        0
```

(17)
```
        6 5
8 ) 5 2 0
  - 4 8
      4 0
    - 4 0
        0
```

(4)
```
    2 2 1
4 ) 8 8 4
  - 8
    0 8
  -   8
      0 4
    -   4
        0
```

(5)
```
        7 6
8 ) 6 0 8
  - 5 6
      4 8
    - 4 8
        0
```

(6)
```
        6 2
7 ) 4 3 4
  - 4 2
      1 4
    - 1 4
        0
```

(18)
```
    1 0 9
5 ) 5 4 5
  - 5
    0 4
  -   0
      4 5
    - 4 5
        0
```

(19)
```
    1 7 3
4 ) 6 9 2
  - 4
    2 9
  - 2 8
      1 2
    - 1 2
        0
```

133

```
(7)    1 5 1    (8)    1 2 3    (9)    1 0 4
    5)7 5 5         5)6 1 5         4)4 1 6
    - 5            - 5            - 4
      2 5            1 1            0 1
    - 2 5          - 1 0          -   0
        0 5            1 5            1 6
    -     5          - 1 5          - 1 6
          0              0              0

(20)    3 0 2   (21)      8 9
     3)9 0 6          5)4 4 5
     - 9             - 4 0
       0 0             4 5
     -   0           - 4 5
         0 6               0
     -     6
           0

(10)    2 2 0   (11)      7 3
     4)8 8 0          7)5 1 1
     - 8              - 4 9
       0 8              2 1
     -   8            - 2 1
         0 0                0
     -     0
           0

(22)      8 7   (23)      7 7
     7)6 0 9          6)4 6 2
     - 5 6            - 4 2
       4 9              4 2
     - 4 9            - 4 2
         0                0

(12)    1 2 4   (13)    3 0 2
     7)8 6 8          3)9 0 6
     - 7              - 9
       1 6              0 0
     - 1 4            -   0
         2 8              0 6
     -   2 8          -     6
             0              0

(24)    1 6 6
     5)8 3 0
     - 5
       3 3
     - 3 0
         3 0
     -   3 0
             0

(14)    1 3 3   (15)    3 2 8
     7)9 3 1          3)9 8 4
     - 7              - 9
       2 3              0 8
     - 2 1            -   6
         2 1              2 4
     -   2 1          -   2 4
             0              0

(16)    1 3 9   (17)    1 5 7
     3)4 1 7          4)6 2 8
     - 3              - 4
       1 1              2 2
     -   9            - 2 0
         2 7              2 8
     -   2 7          -   2 8
             0              0

(18)      7 1   (19)    1 4 0
     8)5 6 8          5)7 0 0
     - 5 6            - 5
         0 8            2 0
     -     8          - 2 0
             0            0 0
                      -     0
                            0
```

Page 90, Item 1:

```
(1)    2 1 7    (2)      8 5    (3)    2 0 9
    3)6 5 1         8)6 8 0         4)8 3 6
    - 6            - 6 4          - 8
      0 5            4 0            0 3
    -   3          - 4 0          -   0
        2 1              0            3 6
    -   2 1                        - 3 6
            0                            0

(4)    1 5 7    (5)    1 4 1    (6)    1 7 0
    4)6 2 8         7)9 8 7         4)6 8 0
    - 4            - 7            - 4
      2 2            2 8            2 8
    - 2 0          - 2 8          - 2 8
        2 8            0 7            0 0
    -   2 8          -   7          -   0
            0            0              0
```

134

(7)
```
      1 0 4
   8 ) 8 3 2
   -  8
      0 3
   -    0
      3 2
   -  3 2
        0
```

(8)
```
      1 2 4
   7 ) 8 6 8
   -  7
      1 6
   -  1 4
        2 8
   -    2 8
          0
```

(9)
```
      2 9 9
   3 ) 8 9 7
   -  6
      2 9
   -  2 7
        2 7
   -    2 7
          0
```

(20)
```
      2 5 1
   3 ) 7 5 3
   -  6
      1 5
   -  1 5
        0 3
   -      3
          0
```

(21)
```
      1 7 8
   3 ) 5 3 4
   -  3
      2 3
   -  2 1
        2 4
   -    2 4
          0
```

(10)
```
      1 1 5
   8 ) 9 2 0
   -  8
      1 2
   -    8
        4 0
   -    4 0
          0
```

(11)
```
        8 9
   5 ) 4 4 5
   -  4 0
        4 5
   -    4 5
          0
```

(22)
```
      1 3 7
   5 ) 6 8 5
   -  5
      1 8
   -  1 5
        3 5
   -    3 5
          0
```

(23)
```
        6 0
   9 ) 5 4 0
   -  5 4
        0 0
   -      0
          0
```

(12)
```
      2 0 2
   3 ) 6 0 6
   -  6
      0 0
   -    0
        0 6
   -      6
          0
```

(13)
```
        9 4
   9 ) 8 4 6
   -  8 1
        3 6
   -    3 6
          0
```

(24)
```
      1 2 3
   7 ) 8 6 1
   -  7
      1 6
   -  1 4
        2 1
   -    2 1
          0
```

(14)
```
        6 9
   9 ) 6 2 1
   -  5 4
        8 1
   -    8 1
          0
```

(15)
```
        9 5
   8 ) 7 6 0
   -  7 2
        4 0
   -    4 0
          0
```

(16)
```
      1 0 9
   9 ) 9 8 1
   -  9
      0 8
   -    0
        8 1
   -    8 1
          0
```

(17)
```
        7 8
   9 ) 7 0 2
   -  6 3
        7 2
   -    7 2
          0
```

(18)
```
        9 4
   6 ) 5 6 4
   -  5 4
        2 4
   -    2 4
          0
```

(19)
```
      1 1 2
   8 ) 8 9 6
   -  8
      0 9
   -    8
        1 6
   -    1 6
          0
```

Page 92, Item 1:
(1)195/88 (2)154/33 (3)33/12 (4)84/40
(5)88/24 (6)62/24 (7)29/14 (8)137/60
(9)90/24 (10)149/42 (11)126/60
(12)96/30 (13)33/12 (14)141/33
(15)65/10 (16)141/35 (17)125/28
(18)48/15 (19)105/45 (20)176/72
(21)144/60 (22)28/8 (23)199/42
(24)118/30 (25)127/14 (26)297/132
(27)291/117 (28)137/60 (29)216/104
(30)154/70

Page 93, Item 1:
(1)66/24 (2)112/21 (3)66/21 (4)36/8
(5)32/6 (6)177/42 (7)296/130 (8)21/6
(9)33/6 (10)85/36 (11)39/12 (12)286/132
(13)42/12 (14)44/12 (15)29/14
(16)175/70 (17)132/45 (18)194/63
(19)61/15 (20)116/21 (21)79/22
(22)206/72 (23)78/30 (24)56/12
(25)40/12 (26)54/10 (27)115/44
(28)81/30 (29)65/15 (30)204/63

Page 94, Item 1:
(1)87/20 (2)202/91 (3)161/42 (4)76/15
(5)241/110 (6)168/60 (7)204/60
(8)172/56 (9)174/36 (10)103/30
(11)97/21 (12)186/55 (13)255/105
(14)115/42 (15)162/42 (16)86/30
(17)231/110 (18)225/70 (19)42/18
(20)299/143 (21)64/28 (22)181/55
(23)122/30 (24)74/10 (25)26/6 (26)66/18
(27)174/45 (28)34/14 (29)135/36
(30)33/12

Page 95, Item 1:
(1)3 12/36 (2)2 25/30 (3)6 1/12 (4)2
45/104 (5)2 79/104 (6)6 7/30 (7)2 11/56
(8)6 4/30 (9)3 61/77 (10)10 3/18 (11)2
9/12 (12)2 6/12 (13)4 3/18 (14)4 17/88

(15)3 58/117 (16)2 75/117 (17)4 43/72
(18)4 1/30 (19)4 14/30 (20)3 66/72

Page 96, Item 1:
(1)8/112 (2)3/20 (3)88/238 (4)3/48
(5)42/156 (6)21/102 (7)10/48 (8)2/18
(9)13/70 (10)13/30 (11)63/234 (12)11/60
(13)1/21 (14)30/136 (15)1/30 (16)47/156
(17)7/16 (18)8/20 (19)4/39 (20)1/12
(21)19/36 (22)34/90 (23)9/143 (24)26/48
(25)14/80 (26)24/72 (27)5/22
(28)115/187 (29)58/112 (30)17/240

Page 97, Item 1:
(1)41/380 (2)4/20 (3)13/285 (4)1/6
(5)9/90 (6)28/130 (7)11/152 (8)10/30
(9)2/90 (10)8/39 (11)56/247 (12)3/60
(13)14/76 (14)6/16 (15)55/136 (16)4/39
(17)21/70 (18)2/56 (19)19/36 (20)49/247
(21)5/24 (22)1/34 (23)107/342 (24)4/15
(25)6/35 (26)8/154 (27)52/90 (28)19/72
(29)2/36 (30)31/209

Page 98, Item 1:
(1)2/117 (2)73/165 (3)112/247 (4)66/208
(5)111/190 (6)6/304 (7)11/18 (8)20/39
(9)62/80 (10)26/68 (11)2/12 (12)12/88
(13)19/40 (14)1/6 (15)8/55 (16)4/30
(17)32/380 (18)7/20 (19)1/6 (20)16/33
(21)7/24 (22)15/144 (23)2/20 (24)19/63
(25)18/76 (26)5/12 (27)10/340
(28)52/240 (29)21/95 (30)1/30

Page 99, Item 1:
(1)62/154 (2)32/66 (3)3/84 (4)3 17/28
(5)6 9/30 (6)3/36 (7)1 10/56 (8)23/30
(9)2 5/18 (10)58/84 (11)7/30 (12)2 6/30
(13)1 58/66 (14)1/6 (15)11/12 (16)5 4/12
(17)31/88 (18)1/12 (19)16/84 (20)46/110

Page 100, Item 1:
(1)30/154 (2)1/156 (3)5/84 (4)4/168
(5)32/50 (6)4/36 (7)12/72 (8)2/8 (9)9/48
(10)2/20 (11)2/30 (12)80/168 (13)21/36
(14)6/80 (15)8/22 (16)32/55 (17)50/156
(18)10/84 (19)6/77 (20)10/90 (21)10/72
(22)15/80 (23)8/88 (24)32/120
(25)12/108 (26)6/120 (27)40/130
(28)3/55 (29)24/42 (30)15/40

Page 101, Item 1:
(1)5/90 (2)3/64 (3)7/228 (4)50/120
(5)28/204 (6)40/108 (7)22/162
(8)100/260 (9)36/117 (10)4/16 (11)8/24
(12)26/140 (13)8/36 (14)12/304
(15)45/340 (16)15/77 (17)27/108
(18)21/120 (19)4/128 (20)18/90
(21)48/180 (22)42/165 (23)96/342
(24)10/36 (25)3/40 (26)24/65 (27)3/16
(28)4/28 (29)4/24 (30)42/320

Page 102, Item 1:
(1)6/51 (2)12/34 (3)28/210 (4)21/187
(5)21/72 (6)32/70 (7)35/255 (8)6/32
(9)6/39 (10)4/72 (11)91/112 (12)154/360
(13)66/300 (14)21/280 (15)20/165
(16)3/26 (17)36/136 (18)8/114
(19)98/198 (20)36/204 (21)36/266
(22)2/54 (23)126/187 (24)2/14
(25)24/143 (26)11/180 (27)132/238
(28)13/266 (29)12/209 (30)28/285

Page 103, Item 1:
(1)1 130/150 (2)1 82/88 (3)1 140/180
(4)16 3/20 (5)7 4/18 (6)4 32/60 (7)2 42/45
(8)7 9/18 (9)1 146/154 (10)5 14/18 (11)5
9/45 (12)14 3/10 (13)2 104/108 (14)2
28/91 (15)3 12/48 (16)1 58/182 (17)7
20/27 (18)4 40/66 (19)1 90/182 (20)3
19/45

Page 104, Item 1:
(1)25/20 (2)25/16 (3)30/27 (4)10/40
(5)56/22 (6)21/48 (7)54/36 (8)36/108
(9)64/40 (10)48/72 (11)16/88 (12)45/63
(13)77/44 (14)44/33 (15)80/30 (16)90/55
(17)12/28 (18)63/18 (19)6/12 (20)28/48
(21)54/24 (22)32/84 (23)28/36 (24)27/32
(25)64/63 (26)56/42 (27)90/72 (28)48/30
(29)36/88 (30)63/16

Page 105, Item 1:
(1)33/110 (2)40/39 (3)22/35 (4)156/117
(5)60/84 (6)24/22 (7)99/126 (8)24/28
(9)36/30 (10)30/70 (11)27/22 (12)156/65
(13)72/156 (14)88/143 (15)36/90
(16)22/40 (17)88/48 (18)49/90 (19)99/36
(20)42/40 (21)154/165 (22)40/48
(23)22/55 (24)132/84 (25)112/84
(26)81/50 (27)60/108 (28)126/70
(29)33/72 (30)54/96

Page 106, Item 1:
(1)99/121 (2)90/112 (3)98/90 (4)132/154
(5)55/45 (6)50/90 (7)32/26 (8)120/156
(9)56/36 (10)36/105 (11)180/135
(12)84/168 (13)99/154 (14)64/60
(15)35/70 (16)36/44 (17)48/60 (18)72/81
(19)49/60 (20)135/60 (21)154/140
(22)120/36 (23)112/90 (24)48/84
(25)117/91 (26)40/63 (27)72/26
(28)168/70 (29)50/20 (30)144/130

Page 107, Item 1:
(1)5 25/28 (2)120/136 (3)1 19/80 (4)30/77
(5)126/255 (6)80/91 (7)78/153 (8)64/119
(9)60/126 (10)3 1/33 (11)42/80 (12)48/66
(13)1 28/32 (14)128/182 (15)2 26/32
(16)7 3/24 (17)171/285 (18)1 29/35 (19)1
30/84 (20)112/117

Page 108, Item 1:
(1)5/7 (2)2/3 (3)7/10 (4)2/3 (5)2/7 (6)5/8
(7)1/3 (8)1/6 (9)1/2 (10)1/3 (11)5/8
(12)5/10 (13)4/9 (14)1/9 (15)3/4 (16)3/4
(17)3/4 (18)2/7 (19)3/4 (20)1/4 (21)1/3
(22)3/5 (23)8/10 (24)5/7 (25)5/8 (26)3/10
(27)2/3 (28)3/10 (29)1/8 (30)1/3

Page 109, Item 1:
(1)1/5 (2)2/3 (3)2/4 (4)1/7 (5)2/4 (6)1/5
(7)3/10 (8)4/8 (9)4/8 (10)1/2 (11)3/5
(12)3/4 (13)5/6 (14)7/8 (15)2/5 (16)2/5
(17)1/5 (18)2/3 (19)3/4 (20)2/10 (21)2/8
(22)5/9 (23)1/2 (24)1/4 (25)4/5 (26)2/3
(27)1/4 (28)1/4 (29)2/3 (30)4/6

Page 110, Item 1:
(1)4/8 (2)2/4 (3)1/2 (4)2/3 (5)2/6 (6)2/5
(7)1/3 (8)8/9 (9)8/10 (10)2/3 (11)5/9
(12)5/6 (13)2/3 (14)4/8 (15)2/8 (16)2/4
(17)3/5 (18)2/8 (19)4/6 (20)3/4 (21)7/8
(22)1/7 (23)6/8 (24)4/5 (25)4/8 (26)1/3
(27)5/7 (28)4/8 (29)1/8 (30)1/10

Page 111, Item 1:
(1)5/10 (2)4/6 (3)2/3 (4)4/8 (5)6/8 (6)1/2
(7)1/4 (8)3/7 (9)7/10 (10)1/5 (11)5/6
(12)4/5 (13)1/3 (14)3/7 (15)2/6 (16)1/3
(17)4/7 (18)1/7 (19)3/8 (20)5/9 (21)1/8
(22)1/6 (23)2/5 (24)5/6 (25)1/2 (26)3/5
(27)4/5 (28)1/2 (29)1/5 (30)4/8

PROGRESS SHEET

	D-1	D-2	D-3	D-4	D-5	D-6	D-7	D-8	D-9	D-10	D-11	D-12	D-13	D-14	D-15	D-16	D-17	D-18
30																		
29																		
28																		
27																		
26																		
25																		
24																		
23																		
22																		
21																		
20																		
19																		
18																		
17																		
16																		
15																		
14																		
13																		
12																		
11																		
10																		
9																		
8																		
7																		
6																		
5																		
4																		
3																		
2																		
1																		

	D-19	D-20	D-21	D-22	D-23	D-24	D-25	D-26	D-27	D-28	D-29	D-30	D-31	D-32	D-33	D-34	D-35	D-36
30																		
29																		
28																		
27																		
26																		
25																		
24																		
23																		
22																		
21																		
20																		
19																		
18																		
17																		
16																		
15																		
14																		
13																		
12																		
11																		
10																		
9																		
8																		
7																		
6																		
5																		
4																		
3																		
2																		
1																		

	D-37	D-38	D-39	D-40	D-41	D-42	D-43	D-44	D-45	D-46	D-47	D-48	D-49	D-50	D-51	D-52	D-53	D-54
30																		
29																		
28																		
27																		
26																		
25																		
24																		
23																		
22																		
21																		
20																		
19																		
18																		
17																		
16																		
15																		
14																		
13																		
12																		
11																		
10																		
9																		
8																		
7																		
6																		
5																		
4																		
3																		
2																		
1																		

	D-55	D-56	D-57	D-58	D-59	D-60	D-61	D-62	D-63	D-64	D-65	D-66	D-67	D-68	D-69	D-70	D-71	D-72
30																		
29																		
28																		
27																		
26																		
25																		
24																		
23																		
22																		
21																		
20																		
19																		
18																		
17																		
16																		
15																		
14																		
13																		
12																		
11																		
10																		
9																		
8																		
7																		
6																		
5																		
4																		
3																		
2																		
1																		

	D-73	D-74	D-75	D-76	D-77	D-78	D-79	D-80	D-81	D-82	D-83	D-84	D-85	D-86	D-87	D-88	D-89	D-90
30																		
29																		
28																		
27																		
26																		
25																		
24																		
23																		
22																		
21																		
20																		
19																		
18																		
17																		
16																		
15																		
14																		
13																		
12																		
11																		
10																		
9																		
8																		
7																		
6																		
5																		
4																		
3																		
2																		
1																		

	D-91	D-92	D-93	D-94	D-95	D-96	D-97	D-98	D-99	D-100							
30																	
29																	
28																	
27																	
26																	
25																	
24																	
23																	
22																	
21																	
20																	
19																	
18																	
17																	
16																	
15																	
14																	
13																	
12																	
11																	
10																	
9																	
8																	
7																	
6																	
5																	
4																	
3																	
2																	
1																	

Be Proud of Your Awesome Work

math award

This certificate is presented to

_____

by

If you enjoy this book please,
write a brief review. Your feedback is
important to us.
Thank you!

Scan the QR code below

Made in the USA
Coppell, TX
20 August 2024